追寻巨人
Seeking the Giant

[美] Jack Sun（予森）- 著◎译

图书在版编目(CIP)数据

追寻巨人 /（美）予森（Sun,J.）著译 . —北京：中央编译出版社，2016.1
（2016.2 重印）
ISBN 978-7-5117-2820-3

Ⅰ. ①追… Ⅱ. ①予… Ⅲ. ①成功心理-通俗读物 Ⅳ. ① B848.4-49

中国版本图书馆 CIP 数据核字（2015）第 259825 号

追寻巨人

出 版 人：	刘明清
出版统筹：	董 巍
责任编辑：	邓永标
责任印制：	尹 珺
出版发行：	中央编译出版社
地　　址：	北京西城区车公庄大街乙 5 号鸿儒大厦 B 座（100044）
电　　话：	（010）52612345（总编室）　（010）52612371（编辑室）
	（010）52612316（发行部）　（010）52612317（网络销售）
	（010）52612346（馆配部）　（010）66509618（读者服务部）
传　　真：	（010）66515838
经　　销：	全国新华书店
印　　刷：	北京时捷印刷有限公司
开　　本：	787 毫米 ×1092 毫米　1/32
字　　数：	120 千字
印　　张：	5.75
版　　次：	2016 年 2 月第 1 版第 2 次印刷
定　　价：	28.00 元
网　　址：	www.cctphome.com　邮　箱：cctp@cctphome.com
新浪微博：	@ 中央编译出版社　微　信：中央编译出版社（ID：cctphome）
淘宝店铺：	中央编译出版社直销店（http://shop108367160.taobao.com）（010）52612349

本社常年法律顾问：北京嘉润律师事务所律师　李敬伟　问小牛
凡有印装质量问题，本社负责调换，电话：010-55626985

目录

推荐语　　　　　　　　　　　　　　　　　　　01
前言　　　　　　　　　　　　　　　　　　　　05
鸣谢与希望　　　　　　　　　　　　　　　　　08
编者的话　　　　　　　　　　　　　　　　　　13
巨人的故事问卷　　　　　　　　　　　　　　　18

第1章：我生命中最黑暗的时刻
　　　　（黑暗边缘的光 - 个人）　　　　　　　23
第2章：浪漫的火花（来自侧面意外的扶持 - 伴侣）　31
第3章："站在巨人的肩膀上"（戏剧化的领导 - 群体）35
第4章：UPS的经历（传奇的领导力文化 - 组织）　38
第5章：急切寻求巨人（成长的需要 - 实践中探索）　44
第6章：沃顿商学院里经历"惰性"
　　　　（意识 - 社会效应到天下）　　　　　　50

目录

第7章：一个巨人的故事（突破 - 转型）   54

第8章：生生不息的巨人传奇

       （实际应用 - 唯美和功效）   71

附录1   期望   81

附录2   我的中国梦   83

参考书目   85

推荐语

很少有书籍能够将细致入微的个人故事演绎为可供他人努力学习的良机。在叙事和隐喻中,予森用勇气和坦诚魔法般地勾勒出他的故事,让我们每个人都可以用自己的方式从中借鉴。他是一个巨人,无论在个人还是职业层面,他的肩膀都可以托起更多成长和更多的幸福感。

——Janet Greco, Ph.D. 宾夕法尼亚大学组织动态硕士项目、沃顿商学院高管教育教授

通过自己谦逊的人生经历，予森向我们显示了我们与自身"巨人"间的关系。这是一个令人信服的互动式故事，它记录了一段追寻意义和重要性的人生旅程。这也是一剂处方，促使我们思考自己的生活、关系以及对他人的影响。巨人会培养明天的巨人！

——Todd Henshaw，沃顿商学院高管领导力项目主任、领导者发展协会主席（加入沃顿前，任美国西点军校领导力项目主任）

"非常高兴——予森出版这样一本简洁但又很有意义的图书,中间的故事真实而不做作,相信读者会非常喜欢。也相信予森未来还能写出更多更好的作品!"

俞敏洪 新东方教育科技集团总裁

一口气读完予森的这本图书,脑海中跳出的第一句话是一位先贤的哲语:"伟人之所以成为伟人,是因为我们跪着。"感谢予森的分享!我们每一个人在学习成长的同时都应该直面自我并努力成为自己内心的巨人!寻找自我,超越自我,也许有一天我们会突然发现自己或许正与巨人比肩!

长啸　中国中央电视台新闻主播

前 言

这本书首先是写给我自己的。书中如实地记载了我追寻自己内心巨人的成长历程。这本书也写给2009年到2012年我在中国人民大学任教时所教过的所有学生，借此，我由衷地感谢与你们共处时我所获得的信任与支持，这段友谊，也是启发我写作本书的缘起之一。希望本书，同样能有助大家找寻到每个人与生俱来的内心巨人。此书，特别也要送给所有期冀在学习、工作和生活中寻求突破成长，在校大学生和刚踏入社会工薪阶层的朋友们。当然，更希望本书能为工作在各个岗位上的领导者、心理咨询师、家长、导师、教师、生活教练，以及我相识和不相识的所有朋友们，提供一个可以和新生代顺畅沟通的共同语言环境。

我也将这本书送给我的两个儿子：孙乐星和孙星宝。希望日后，这本书能有助于你们找到适合自己的人生旅途方向，帮助你们认识一个和平常有点不一样的父亲。

我出生于中国古都西安附近的农村地区，并在那里长大，是普普通通的农民后代的一员。即使现在闭上眼睛，小学一年级时坐过的那个石凳，我依然能感觉到它的冰凉。还有覆在寒冷教室窗户上像纸一样薄的破塑料膜，依然清晰在目。这是家乡小学第一个冬天留给我永久的记忆。

命运却总有其他的安排。经过许多偶然却又似乎平常的生活事件，二十多年后，我最终辗转地来到了美国生活和工作，并就职于全球顶尖的沃顿商学院。有人说：你已经成功地实现了你的美国梦——对吗？理智和清醒下来，我扪心思考，我的确实现了一个中国最底层农家孩子没有做过的梦想。但是，我所经历的生活故事，却又是"梦想"二字完全不可能描绘出平凡而又非常的故事。如果我可以用一句话来概括我至今的生命历程,我会总结为:一个并非平凡的领导力探索和学习的历程。

早在2003年,在我前往美国生活和学习前,我当时的一个同事对我说了一句话:"Jack,你马上要站在巨人的肩膀上了,你一定要确保充分利用这样一个难得的机会。"当时我不能完全理解这句话。但就在不久的未来,我不但深刻地理会了这句话,而且这句话也成了我为人处事的一个启示和标杆。

感恩我生命中遇到的所有长辈、导师、挚友和生活教练们,正是你们的慷慨帮助和激励,唤起我内在的觉醒,助力我找到自己的人生财富,并得以分享、帮助并利益他人。

作者

鸣谢与希望

我的爱妻 Cherie Sun 从一开始就一直支持我写这本书，她挚真情感上的支持使得这一切成为可能；两个儿子乐星和星宝为此牺牲许多和我在一起玩耍的时间，尤其感动的是他们还帮这本书画了插图；感谢我父母的养育之恩以及岳父母 George and Carol Benner 对我们家庭和孩子的一贯支持。忠诚地感谢我的所有亲人们。

当然，还要由衷地感谢所有在我成长过程中曾经影响过我的老师们、导师们和生活教练们。感谢 Steve Shepard 教授对此书所提的忠实诚恳的意见，感谢苏江、方菱、刘峰、Joern Geisselmann、Margaret Baumann、Edmond Baumann、Tony Heath、Karen Hopkins、

Rebekah Zanders、Steve Zanders、Peter Scott、James Thomason、Todd Henshaw、Sue Wharton、Paul Wharton、Deb Giffen、Sumathi Pearl、Adam Grant、Anne Corcoran、John Percival、Mike Useem 等等，以及所有那些在我的内心留下了印记的人们。

还要感谢好友们对我的书稿所提出的忠肯建议，他们是：牟玉梅、赵生伟、王巍峰、丁超、Jessica Kotain、Hilary Zee。

费城免费图书馆和 Overdrive 应用程序，也是我要感谢的。正因为有了免费的图书馆和现代技术的完美结合，我才得以访问并聆听了无数关于畅销的领导力的音频书籍。

最后，由衷感谢我有幸在沃顿商学院听过的所有关于领导力的课程，如果不是借助沃顿商学院诸多知名教授的知识和启发，我也无法成就这本书。更至关重要是，我必须向宾夕法尼亚大学执教的 Janet Greco 教授致以特殊的感谢。她是沃顿的教授之一，同时任教于组织动态硕士课程：DYNM673：组织中的故事：执行发展工具。

这个课程是我选修过的最实用和有益的一门课程。她的教学启发了我去收集所有关于我的"巨人"的故事。如果不是因为她在第一堂课上展示她对故事如此美好的愿景，我也无法受到启迪写出《追寻巨人》这本书。我根据她的愿景勾勒了下图，与大家分享：

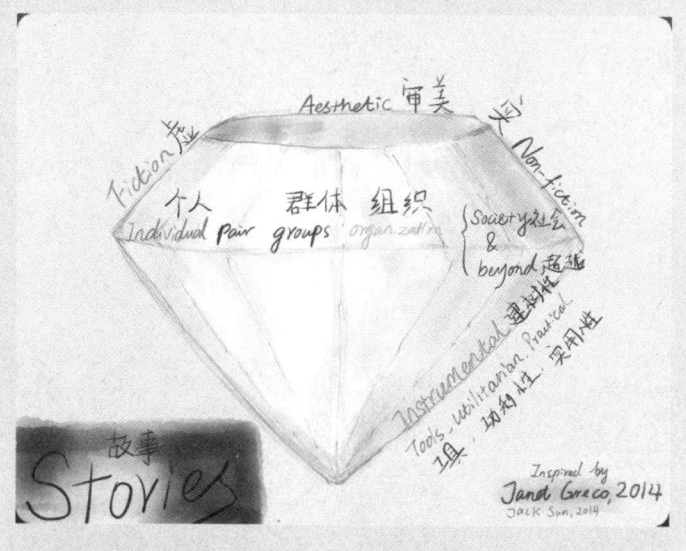

她的这个关于故事的愿景具有极强的系统性和建树性，可以看出她对故事的愿景在这美丽的钻型图里被表

现得淋漓尽致。我们每个人、每个家庭或团体，还有每个组织都有自己独一无二的故事。不管他们的故事是真实还是虚构，只要把各类故事巧妙地结合起来并作为服务社会的一个良好的工具，那么，这些故事就是美好的。更重要的是，如果有建树地使用这些故事，将对我们个人、团体、组织和社会，乃至于整个世界都会产生无法估量的积极影响。由此，根据 Janet Greco 教授的这个愿景，我开始了本书的写作。

在听 Janet Greco 教授的《组织中的故事：执行发展工具》这个课程之前，我原本是蛮排斥听人讲所谓的"故事"的。也许和人生经历以及态度有关，潜意识里我经常不自觉把"故事"俩字，和"胡编乱造、流言蜚语、虚伪不实"等词联系到一起，因为我最讨厌这类东西。我太太有时会读一些虚构的小说之类的书，她可能会在读的时候被感动得流泪，我觉得不可理喻。我对她说，"我真不明白你为什么会浪费自己宝贵的时间去读别人虚构的故事，而且你还为那些虚构的故事而浪费你的感情？"曾经，对听别人讲故事那么不屑一顾的我，直到遇到 Janet Greco 教授把她的故事课程的美丽愿景展现在我面前时,我突然意识到:我之所以厌恶故事是因为我以前把所有的注意力全放到故事的消极方面或是消极的故事方面了,她这美丽的故事愿景让我忽然看到了故事积极的一面,我才真正地意识到了有价值的好故事可以对我们个人、团体、组织以及社会带来难以想象的实用价值和如此不可估量的积极效应。我为此前的认知和思维方式反省,很惭愧地说,Janet Greco 教授的第一节课就改变了我,所以,于我而言,在这本书里你也将会读到一些非常有价值的故事,希望这些故事能启迪你发现自己故事里宝贵的价值。而且,我也希望每个读者在明确意识到自己宝贵价值的基础上,将这些价值勇敢地体现到自己的人生故事中,并将那些不可估量的价值传承下去。

作者

编者的话

这是一部帮助我们探寻内心巨人、激发内在潜力的心理学作品。在本书中,作者穿插本人一系列真实的故事和实践调查,并结合了心理学、社会心理学、结构组织学和领导力等前沿知识,深入浅出地启示每位读者寻找各自独有内心的巨人。在读这本书之前,明确以下几个单词的定义概念会有助读者更好地理解作者的用意。

价值(维基百科):http://zh.wikipedia.org/wiki/价值

价值,泛指客体对于主体表现出来的积极意义和有用性。可视为是能够公正且适当反映商品、服务或金钱等值的总额。

在经济学中,价值是商品的一个重要性质,它代表该商品在交换中能够交换到其他商品的多少,价值通常通过货币来衡量,成为价格。这种观点中的价值,其实是交换价值的表现。

本书里提到的价值是除去经济价值之外的价值。如果我们把经济价值称作硬价值,那么我可以把这本书里提到的价值称为软价值(http://www.baike.com/wiki/软价值)。所谓软价值是一种无形的心理价值。所以,在这本书中,一般情况下你所看到的价值都是这种无形的软价值。爱因斯坦一句名言是这样说的:不是所有有价值的东西都可以用经济价值来衡量的,同样,不是所有经济价值的东西值得被衡量。(Not everything that counts can be counted, and not everything that can be counted counts.)爱因斯坦的话辩证地告诉了我们软价值虽然不能用数字之类的标尺来衡量,但这种软价值却在很多情况下是值得我们用心去衡量的。

巨人（维基百科）：http://zh.wikipedia.org/wiki/巨人

巨人，是神话、传说或童话中常见的生物，几乎全世界都有他们的踪迹，由希腊神话、印欧语系神话，到中东、亚洲及美洲地区的神话及《圣经》内的故事都有他们的痕迹。而现在的社会中，巨人依然出现在神怪电影或电玩游戏里。这些高大异常的生物，尽管有时被描写成温和而无害，但大部分的故事都是将他们描述成可怕、贪婪、食人且又愚昧的怪物，除了拥有惊人的力气与食量外，还以攻击行为、战争等方面的事迹闻名。

本书中的巨人是用比喻的方式把无形的软价值拟人化。也就是说，这个"巨人"和维基百科里所指的巨人大相径庭。牛顿曾经说过：如果说我看得比别人更远些，那是因为我站在巨人的肩膀上。（**If I have been able to see further, it was only because I stood on the shoulders of giants.**）所以，这本书邀请读者和我一起追寻和造就一个属于你自己独一无二的内心巨人。

励志：http://zhidao.baidu.com/question/254390869.html

励志是一门学问，这门学问不管多牛的人都读不懂、学不精，进而形成一个独立学科——"励志学"。励志学，不仅仅是要激活一个人的财富欲望，更要激活一个人的生命能量，唤醒一个民族的创造热情。失去创造力，是一个人乃至一个民族最大的悲哀。而励志，便是让一个人重新焕发起这种力量。励志，并不是让弱者取代另一个人成为强者，而是让一个弱者能与强者比肩，拥有实力相当的生命力和创造力。励志，即是唤醒一个人的内在创造力。唯有从内心深处展开的力量，用心灵体验总结出的精华，才是一个人真正获得尊严和自信的途径。

本书没有励志类标签，没有发财致富的诀窍或捷径，"通过审视我们自己内在的软价值，从而激活我们内心深处正能量的源泉"是本书最大的特点。

希望本书在帮助读者探寻内心巨人的同时，也能激发读者的领导潜力,激励我们通过自觉和以身作则,给我们生存的这个世界和社会奉献出来自于我们内心巨人所创造出的、金钱所无法衡量的价值。

编者

巨人的故事问卷

在这本书中,我邀请每一位读者通过交互参与的方式来主动完成这个巨人的故事问卷。通过你的亲历经验,我希望你能体验到这个故事问卷的神奇之处。因为它很可能会打开你智慧的月光宝盒。同时,在这本书的每一章也附加了巨人问卷的额外练习,以进一步增强这本书的实用价值。如果读者有任何疑问请发电子邮件到:*seekingthegiant@yahoo.com*,或者访问新浪微博:追寻巨人。

在你继续阅读这本书前,让我们先完成以下问卷。回答下列问题将会帮助你更好地理解和受益于这本书。

1. 假设你站到了一个巨人的肩膀上,你会看到什么,你又感觉到什么?充分发挥你的想象力,这个巨人可以高达任何你可以想象的高度。根据你的理解,可以写下或画出你的答案。

2. 既然你站到了这个巨人的肩膀上,也就意味着你和巨人之间有了一定程度的信任。所以,请你列出3到5个形容词来表达这个巨人的性格特点。

3. 如果你站在肩膀上的那个巨人跌倒了,你会怎么做?(这是情景问题,巨人跌倒了,他可能无法再站立。写下在此种情景下你会怎么做。)

4. 如果你看到另一个巨人行走在你现在站在肩上的巨人附近,而另外一个巨人要比现在的这个巨人高大得多,在此种境况下,你会做些什么?

5. 如果你站在巨人的肩膀上太长时间后，想象会发生什么？

完成以上《巨人的故事问卷》之后，我很荣幸欢迎每位读者与我一起分享我的追寻巨人之旅。真诚地希望我在这次旅程中的每次会心一笑和收获，都能启示各位读者，而有助于激发和开启属于每个人自己的、那独一无二的巨人探索之旅。

第1章

我生命中最黑暗的时刻
（黑暗边缘的光-个人）

出生在20世纪80年代的中国人，被认为是非常幸运的一代。那时候的中国在经历了战乱、自然灾害、政治运动之后，开始了改革开放，人民的生活条件正逐步改善，我和我的同龄人恰是赶上了这个时代出生的幸运儿。当时，从主流媒体看到的信息是，中国这条巨龙（沉睡的巨人）从睡梦中醒来了。像全国人民一样，我们满怀希望地期盼自己在这个"觉醒的巨人"的看护下也一起觉醒。至少，新闻和电视是这么告诉我们的。然而，当谈到我自己个人的觉醒时，这件事却绝非必然，而且是另有渊源的。

对于20世纪90年代大多数的高中学生上大学就像

买彩票中大奖一样难。因为当时的国有高校录取率只有3%-7%。我在第一年高考落榜后，硬着头皮在复读那年幸运地挤进了大学的门坎。我考上大学"纯粹是运气。"这是我当时的一个高中老师的原话，我清楚地记得接到录取通知书的那天，全村人似乎都把我当成一个名人。在此之前，当我走在街上，没有人会察觉到我。但是，在我考上大学之后，走在自家附近的街上，人们开始用充满钦佩和羡慕的眼光关注我了。当时，我感觉自己好像已经征服了整个世界。

走在云端的悠悠忽忽的状态并没有持续太久，当大学注册入学时间渐渐逼近时，我和家人都不得不面对一笔可观的我们无法负担的学费。当时，大多数农民并没有多少积蓄，更不懂得如何从银行贷款之类的事。我的父母也只有从亲戚朋友那里借钱为我支付学费。我永远不会忘记母亲走出舅舅家门的那一天。当时我和母亲去舅舅家借钱。如果他愿意的话，他可以借给我母亲所有我应缴的学费。我不记得妈妈对舅舅说了什么，只记得舅舅说："这里是200元，你拿着，不用还给我了。"我母亲很尴尬地接过那200元钱。我以前从未见过她如此哑口无语,她失落地示意我走出了舅舅的家门。当她迈出舅舅门外的那一瞬间,我看见她眼睛里充盈着泪水,一串泪珠从她脸的一侧情不自禁地滑下。我无法想象当时她有多么难受和苦楚,因为我的父母在舅舅困难时曾经全力地扶助过他。我母亲有8个兄弟姐妹,因为她是最年长的姐姐,所以在母亲小学五年级时,不得不被迫辍学帮外公外婆操持家务并照顾弟弟妹妹。那天晚些时候她对我说,她不是一个好姐姐,年轻时没有教导好自己弟弟。她还告诉我,我们给予别人的一定要多于向别人索取的,因为只有这样我们才可以坦荡荡、心平气和地过日子。那一天,我把母亲的话深深地刻到了心里。我也知道她说那句话的那一刻,她已经原谅舅舅了。

看着父母千辛万苦地给我借到足够的学费,实在是于心不忍。于是,我暗自许下承诺一定会努力学习,并找一份好工作帮父母还清所有他们为我读书欠下的债务。我怀着要为家里还债的信念走进神秘的大学。

对我来说,没有花太长的时间我便意识到了现实中的大学不是我想象中的那样浪漫无邪。首先，我意识到，作为一个农民的孩子适应在城市里上大学是非常困难的。贫富悬殊和生活方式都很难融合。城市富足的生活方式让很多像我一样的农家子弟心理失去了平衡。很多时候，我会耳闻到一些对乡下人的风趣嬉笑。时常听到最多的是"你这是小农意识""你真农民"等。虽然大多是半开玩笑的话语，但作为一个农民后代，我的心里还是明显地受到此类歧视的影响。为了融入城市的社交圈，我的周围甚至有从乡下来的学生会假装他们是从城市来的。渐渐地，我也开始纠结我自己的身份并苦苦挣扎在价值观濒临崩溃的危机里。有些同学告诉我，我太老实了，笑得太多。他们告诉我，如果我酷一点，就会吸引很多女孩的注意。在价值观和信仰的撞击下我逐渐地迷失了自己。有一段时间，我假装自己很酷，试着撒谎，掩盖自己，脆弱、郁闷、酒精、色情、自杀意图充斥我迷失的灵魂。我无力地挣扎在这些破坏性的行为里。我也试图追求过周围的几个女孩，但结果都不能逃脱这样一个刻板老套的故事：一个无望的贫苦农民的孩子想约会城市女孩，真是癞蛤蟆想吃天鹅肉、痴心妄想。在我的内心世界混乱之中，我也无法专注于学习。黑暗、抑郁几乎是我大学前两年唯一的记忆。无论我怎样努力挣扎，还是没有看到任何希望，也没有找到任何生命的意义。我拼命地在绝望中挣扎。每天我试图把强装的微笑挂在自己的脸上，假装一切都正常。然而，内心的包袱是如此沉重，我甚至不知道我是谁了。明显地感觉到自己就像被卡在最黑暗的隧道里，努力地喊救命。但无论自己内心如何呐喊，似乎没有任何人在那里聆听。

在我大三的时候，作为一个学英语的学生，我的学习负担变得比以往任何时候都重。这在某种程度上使我暂时忽略了我不安的内心世界。但是，我的内心仍然像一个定时炸弹，可能随时爆炸。

一个很偶然的机会，我认识了隔壁学校读中文的一个德国朋友。在开始的时候我们互惠互利，他帮我练习英语，我帮他练习汉语。渐渐地，我逐渐认识到这个德国人一点儿也不像以前认识的任何外国人。当时，作为一个无神论的我开始怀疑"人确实有灵魂"。这个德国人的神秘显然唤起了我对有关宇宙之谜疯狂的想象。那时候,我一点也没预料到他将是我一生中最伟大的导师之一。有一天,我们坐在初秋校园的曙光下做完语言练习。我鼓足了所有的勇气,向他倾诉了那一段时间所有的纠结和对生活的挣扎,包括我内心深处最黑暗的秘密。我当时很害怕,因为我担心他知道我内心如此黑暗后,可能不打算继续做我的朋友了。但是,他之后所说的话却成了我以后反复告诉给其他朋友们一个经典的启迪性故事。他当时对我的回答是:"Jack,诚实、正直、善良、纯粹是你内在的价值,这些价值没有任何人可以从你那里抢走或是偷走。除非你自己放弃。因为它们是天堂的宝藏。如果你选择,这些天堂的宝藏将永远和你在一起。"("Jack, honesty, integrity, kindness and truthfulness are things nobody can rob away from you unless you give them up on your own. They are heavenly treasures that will stay with you forever if you choose to.")

应用练习

• 你有没有经历过心灰意冷的时候?当时,是谁或者是什么引导你走出那些艰难的时光?

• 珍藏在你内心里的天堂财富是什么?

This is my heavenly treasure

第 2 章

浪漫的火花
（来自侧面意外的扶持 - 伴侣）

一年半以来，我的德国朋友、导师从精神方面一直精心地辅导我。得益于他的帮助，即使最黑暗的乌云也不能阻止我看到乌云之上的蓝天了。我由衷地感激这个德国导师的鼓励。因为他，我不仅走出心理的抑郁，而且在情绪上和精神上也渐渐地成熟起来。慢慢地，我开始体会到作为一个完整的人的充实和喜悦。在他的引导下，我最终成功地进入了所谓的自我创作心态（self-authoring mind）（Kegan & Lahey，2009.）。说实话，我在毕业那年，虽然还仍是一个孤独和贫困农村的孩子，但是，大四那年是我最快乐、最心满意足的一段时光。另一个我预想不到的结果是，多年后一个大学同学告诉我说:"我经常告诉我的学生,我以前大学时的班长 Jack 是多么棒的一个班长。你总是那么积极向上,而且,总是那么关心我们。"听到这位同学的话,当时真是令我感到诧异,因为真没想到当时的同学是这样高看我(如果她们真的知道大学前两年的我,可能就另眼相看了吧)。无论如何,这也让我意识到:我导师的话已经大大改变了我的命运,我的改变甚至已经影响周围的人。这可能是我第一次感受到站在巨人(我的导师)的肩膀上会给我的命运带来如此大的改变。

当时,浪漫对我来说从来都是一个很陌生的字眼,我也完全不理解浪漫的含义。直到我在大四那年毕业前遇见了我现在的妻子,一切都变了。我没有一见钟情坠入爱河,不过,当她第二次踏入我视线的记忆,到现在那一幕的画面仍旧像看超清电视一样鲜亮清晰:那时,她和她的父亲在我们学校的操场上看我打篮球,她盘起了她那漂亮的金色卷发,加上运动休闲装扮,整个人都是那样的清雅脱俗。不知道为什么,第一次在教室里看到她的时候,却没有此般感受。但在再次见她的这一刻,我的脑海中闪过了她的整个生活画面：她出生时的婴儿啼哭；当她六七岁时脸上天真幸福的笑容；现在站在我面前的这个华丽而可爱的天使；最后是当她变成一个老太太头发花白，但仍然在她眼里泛着恩爱的样子。那一刻，我感到一种从未有过的宁静感与平和。——"就是她了，她是我命中注定的另一半了。"十四年后，我们仍然一起编写着这个美丽而浪漫的故事，我真不知道怎么感谢上天，感谢它给我这样一位美丽动人的女人。在我们结婚不久，她说的一句话仍然深深地印在我的脑海里："杰克，你是我们家的一家之主。我和我们的孩子现在和未来都会跟随你。我只想让你知道，你的决定将是我们的决定。即使我不总是同意你的决定，但只要你坚持，我们仍然会跟随你的。"听完她所说的话，我对自己说"什么，即使你不同意我的决定，你和我们的孩子仍然要跟着我？这是一个什么样子的逻辑？！"正是我的妻子、我的生活伴侣的那句话，在那个时候很给力地推了我一把。她不但鼓励我更好地做人，而且也把我推到我们家庭的领导者地位上。正是这种信任和托付，推动了我不懈努力地做一个好男人和好的家庭领袖。

应用练习

· 在你的记忆里,有没有人曾经授权给你,并让你感觉自己得以提升?如果有的话,请将你的经历写下来,并和周围关心你的人分享一下你的独特经历。

· 请在网络上观看这个关于领导力的 TedTalk 视频,"First Follower Leadership Lessons from Dancing Guy 第一个跟随舞者跳舞的人对领导力的启事"。然后,反思一下,是什么造就了领导者:http://ed.ted.com/on/IgslePtt

第3章

"站在巨人的肩膀上"
（戏剧化的领导－群体）

我毕业不久就搬到了北京，并找到了第一份正式工作，就职在一个英语培训中心做学生辅导员。我在培训中心工作直到它被关闭后的七个月左右。在获知培训中心将被迫关闭的两个月前，当时的CEO已经停止支付我们员工工资，大部分员工离开了公司并寻求新的机遇，有些员工甚至光明正大地带走了一些公司的昂贵办公物品，以此来抵消他们被拖欠的工资。最后，仍留在中心工作的只剩三四个人，其中之一是培训中心的教务校长，她是我所见过的最善良的女人。她曾对留下的人说："我不是商人，我知道无法改变CEO关闭中心的决定，但我是一个教育工作者，我能做的就仅仅是传播知识给学生，帮助他们学有所成。我相信你们自愿留下，是因为你们也都相信教育是我们唯一的宗旨。只要还有一个学生在我们中心学习，我就会继续呆在这里，帮助这个学生学习，直到中心被关闭的最后一分钟。"

这也许是我人生经历中精神最崇高富足的一段时间，在两个月的薪水没有被支付的情况下，我和校长，还有其他几个同事坚守到培训中心被迫关闭的最后一刻。也就是最后那一天，在我们培训中心学习英语受益的众多学生中，有一位女士眼里含着泪水激动地告诉我们："我今年55岁了，从来没有想过我能学好英语，是你们这个培训中心给了我希望，并帮助我获得了学习英语的信心。我在这里学习了仅仅5个月，已经学会了用英文进行日常对话。我知道我在美国的女儿一定会感到如释重负了，因为，凭我现在的语言技能已经足可以自己生存了。对此，我非常感谢你们所做的所有的辛勤工作。"她的一番感激话是如此平淡，但却让我感到舒适和愉快。是的，就在那一刻，拖欠了几个月的工资和曾经发的牢骚，似乎没有了任何意义，显得那样微不足道。

这位教务校长，就是曾经跟我说"Jack，你马上要站在巨人的肩膀上了，你一定要确保充分利用这样一个难得的机会"的那位智者。通过这次经历，我和她成了忘年交。多年以后再回顾那段经历，她显然是我人生中遇见的又一巨人"导师"。她对教育的热情以及无私奉献精神一直在潜移默化地感染着我。也就是从那时候起，我开始认识到：将来要成为一个什么样的领导者，是由自己的选择来决定的。像我的这位校长朋友，她选择成为一个授权给他人的领导者，即使牺牲了自己的利益，她依然会无私地托起和帮助她周围的人。

实用练习

• 搜寻在你的生活中，你曾经遇到过的"巨人"。写出这些"巨人"的名字和他们身上的个性特点。

• 写下这些"巨人"们是如何在个人层面上影响你的那些故事片断。

第 4 章

UPS 的经历
（传奇的领导力文化－组织）

不难得出结论，我的母亲、我的德国心灵导师、我的妻子和我的校长朋友都是我生命中幸遇见过的巨人。他们都或多或少地影响和塑造了我的品性和价值观。而且，他们也辅助了我领导能力的提升。这些都是我生活中刻意选择记住的一些故事和情节。也就是这些巨人，和他们的故事渐渐地成就了今天的我。我现在清楚地看到我从他们身上学到了这些集体特质：宽恕、诚实、真诚、信任和授权。

早在 2006 年，在我被聘为 UPS 销售代表的那段经历可能是迄今为止我经历过最具挑战性的工作之一。且不说当时我刚上岗时要学习的所有技术和企业知识，再加上文化和语言障碍,这对我在 UPS 的工作带来了很多难度。记得当时做销售时,我要面对各种心情的客户,还有各种要求的客户。这对任何一个美国本土的销售员来说都是一种挑战。但是,正是这些经历挑战了我的极限。由此,我在 UPS 的那段时间里迅速地成长着。UPS 在各个方面都被视为商界的巨人,比如它的领导力、文化、产业、传奇、全球化、技术等等,都是走在世界的最前沿。我从第一天在 UPS 工作开始,就如饥似渴地将自己泡在知识海洋中,不断地学习和成长。几个月的培训后,我的一个领导告诉我:"杰克,你将成为 UPS 的品牌,你站在广大客户面前就要代表 UPS。所以,现在是你大显身手的时候了!"这些话是如此的有分量和激励,即使面对一切逆境,我当时觉得自己像一个超人一样地无所畏惧。回首那段经历,面对那些所有的冷漠陌生人的电话,所有不开心以及愤怒的客户,我能够将最好的 UPS 品牌,以最敬业的姿态呈现给所有的客户。我认为这完全取决于自己相信的那句话——"我就是 UPS 品牌。"所经历过的 UPS 文化,是体验过的最好的组织文化之一。告诉我那一番话的UPS领导，是我最敬佩的领导者之一。记得还有一次，在我的努力下帮助团队获得了和东海岸最大的一个电力设备供应商公司召开一个业务洽谈会议的机会。我已经安排了两个UPS物流专家和我的经理一起参加，同时，该公司的CEO、CFO、律师和首席物流官也都应邀出席这个会议。让我万万没预料到的是，会议开始前五分钟，经理对我说："杰克，正是由于你的努力，才能与该客户重新建立业务对话关系。我相信你比任何人都了解这个客户。你不介意主持这个会议吧？"我报以绝对怀疑的态度看着他，"你真的认为我可以主持这次会议吗？"我犹豫了一下。"打起精神来，杰克，我们知道你的业务能力是多么精湛，我们绝对相信你！"他坚定地看着我。我实在无法抗拒那一刻他给予我的信任。直至今日甚至不知道自己是如何在那次会议上鼓起勇气和自信的，不仅在会议上成功地展现了UPS的巨人风范，还帮助UPS团队拿下了我的第一个百万美元合同——我同样要感恩这位经理。

考虑到孩子的汉语教育问题，我和妻子决定回到中国发展。这样，我们的孩子就能够更好地学习中文。我依依不舍地告别了UPS。我仍然无法忘记2008年底和UPS同事说再见的那一天，为表彰我为UPS服务一年半的时间，主管给我颁发了这一生至今唯一得到过的一个牌匾，并让我给大家一个正式的辞别，这样我的同事都知道我人生旅程会走向何方。那是一个我经历过的最棒的告别会。部门主管告诉我，一般的员工离开UPS后，人力资源部门会关闭雇员的应聘记录，也就是说再也不能被应聘回来。不过，她表示，UPS将保持我的应聘状态为打开。这就意味着，如果我将来想再回UPS，他们的大门会为我敞开。就这样我结束了我的UPS旅程，并在UPS画了一个圆满的句号。在那之后，我不得不承认，即使离开UPS这么多年了，我还是觉得身体的一部分还在流"UPS棕色"的血液。传奇的UPS公司教给我传奇的领导风范。我知道任何语言都无以表达我对UPS以及那些可敬领导者的感激之情。因为，他们留在我内心的不只是一个传奇而已，那无疑是我受益终生的重要精神财富。

应用练习

·如果你把你工作的单位看作是一个巨人的话,这个"巨人"的个人性格特点是什么样子的?

·如果这个巨人、组织让你失望了,你会如何反应,或者你会怎么做?

第 5 章

急切寻求巨人
（成长的需要 - 实践中探索）

那是一个星期五的早晨，我急冲冲地赶回上上午 11 点的课。和预期的一样，我迟到 30 分钟。当走进教室并示意我妻子，她可以离开了（她是我学生的 30 分钟的替补老师）。我的妻子离开后，我的学生明显有点混乱不安。终于，其中的一个学生问道："老师，您今天为什么穿了西装，而且迟到了半个小时上课？""我去面试了。"我回答道。"您的意思是，您可能不再教我们了？！"那位学生的脸明显无法遮掩他的震惊。"是的，我想问一下你们可以告诉我一个应该留下来教你们的理由吗？"我终于向我的学生坦白了自己内心的纠结："当我给你们布置了课后作业，你们不做；当我在课堂上提问，你们没有回答；当我鼓励你们在课堂上要多提问题，你们没有人问……我觉得我是一个很失败的老师，因为我无法与学生互动和交流。""不是这样子的，老师！我们很喜欢您的课。您是我们所有的老师中最好的一个。"一个学生提高了嗓门。"我不理解。如果你喜欢我的课，那为什么你们没有积极地参与课堂内外的活动，为什么你们也从来不问我任何问题？"我说出了自己的困扰。课堂突然沉寂了大概有30秒，但却感觉像一个小时。"老师，我们有问题，我们想问你很多问题。但是，我想只是害怕在公众面前问问题。"一位学生喃喃自语道："我们只是没有自信。因为，我们大多数的英语基础都比较弱，我们都担心在两年内我们无法学好英语然后出国学习。""你们怕什么？难道我看起来像一只老虎，会咬人吗？我不得不承认我真的不了解你们。我觉得，我也不是一个好老师。因为我甚至都不知道你们在想些什么。我很高兴今天我们终于开始真心交流了。"

就是2009年的那个春天，我几乎放弃了教书。在此之后，我特意安排和所有的学生进行一次面对面的谈话或咨询，这样有助于我去了解他们。通过与每个学生的交流，我发现了一个全新的世界。当时我每学期会教150名左右的学生。鉴于人多，我让他们可以两三个同学在一起。如果有特殊情况，可以一对一。当我们开始互相交流时，渐渐地两个世界的隔阂消失了。经过多次面对面促膝交谈我才知道，我的学生基本上都是出生在经济基础富裕的家庭。他们的父母要么是富商要么是政府官员。"富二代"或"官二代"是当时社会给他们的绰号。然而，几乎所有的学生都不喜欢这些称号，"富二代"或"官二代"明显地被社会附加了一层贬义。我忽然发现，他们的大学生活和我的大学生活有种说不出来的某种相似。我当时被贴上的是"农民""乡里人"的标签。虽然"富二代"或"官二代"和"农民""乡里人"从经济程度上是两个极端，但是，有趣的是这些标签或绰号都远远不能说明，也不能定义我们是谁。我们内在本我的价值远远大于这个社会和他人给我们的定义。

通过我和学生之间的非典型友情，我发现，在我们之间是没有界限的。因为在内心深处，我们都渴望被尊重，我们的灵魂都在寻求高尚存在的意义。

当时，很幸运的是，我的部门主管完全信任并授权给我去教这批学生。她对我说："我知道你有丰富的国外学习和工作的经历，而且，我非常相信你的能力。所以，这些学生的学习和成长就完全托付给你了。"在接下来的三年里，每学期我会根据学生的具体情况制定教学计划。初期，我根据学生及时的反馈，每两至三个星期会修改一次教学计划。后两年，为了更好地迎合多数学生的具体情况，一学期我会修改教学计划三四次。当时，我和家人正好住在大学校园里。所以，通常情况下，晚上8点以后，我的时间表是为学生开放的。有时，我们会有领导力学习小组，有时是一对一的辅导或心理咨询。我贤惠的妻子意识到学生是多么需要帮助，我的工作也有幸得到了她的大力支持。她决定晚上8点后是最佳时间去帮我的学生，因为我们的孩子每晚8点准时睡觉。我算是使尽了浑身解数去激励、教育、引导或指导这些学生。有几次，我不得不给学生作心理咨询直到午夜；有几次，我不得不立即飞到另一个城市，进行危机干预；有几次，我发现自己在帮助学生抵御抑郁症和其他心理问题时到了山穷水尽的地步；有几次，我不得放弃自己的所学而向上天投降……所有这些经历都让我更加渴求知识和自己的内心成长。我意识到了自己的能力是多么有限，在所有我关心和想要帮助的人面前，我时常显得是那样的无力和黔驴技穷。就在那时，我学会了向我的信仰、家庭和友人寻求情感和精神支持。因为任凭我个人的能力根本无法帮助学生。也就是那时候，我意识到了精神财富对我们任何人来说是多么的至关重要。

在那三年期间我很意外地和几百个"富二代"或"官二代"学生打成一片。我从他们那里听到了很多让人心痛纠结的故事。在这群可爱的学生面前，我虽有一腔热血，却孤军难敌。我无意中读到《纽约时报》2011年转载的一篇博文："中国啊，请放慢你飞奔的脚步，等一等你的人民，等一等你的灵魂，等一等你的道德，等一等你的良知！不要让列车脱轨，不要让桥梁坍塌，不要让道路成为陷阱，不要让房屋成为废墟。慢点走，让每一个生命，都享有自由和尊严；让每一个个体，都不被这个时代抛弃。"就像我的学生一样,我发现我们都在道德退化、社会和经济迅速发展与变革的巨大浪潮面前显得如此的无能为力。作为 20 世纪 80 年代的一员,我亲身经历了这种剧烈的变化,并亲眼目睹了金钱和物欲是如何折磨与扭曲着人类脆弱的灵魂。像许多中国人一样,我非常深情地关注着祖国的未来,并希望她可持续性地发展。我们都知道中国作为东方的经济巨人是站起来了,但我相信中国作为一个精神上的巨人也将挺拔地站起来。

应用练习

•当你发现自己在无奈和绝望的时候,你会从哪里寻求帮助?你是如何摆脱瓶颈境况的?

•试想一下,如果你有一个永远在你身边并永远不会让你失望的巨人,他会给你的生活带来怎样的经历?

第6章

沃顿商学院里经历"惰性"
（意识－社会效应到天下）

为了孩子的教育，我和家人又回到了美国。我们孩子所在的一所北京国际学校，由于一个无法控制的原因终止了奖学金。此外，作为一家之主，我也意识到，真的需要慎重地考虑自己的职业生涯了。所以，全家恋恋不舍地回到了美国。我怀着对高等教育坚定不移的激情，几乎在所有东海岸的院校都递交了简历。幸运的是，沃顿商学院应聘了我项目协调员的职位。

当我开始在沃顿商学院工作时，很快就意识到，再一次站在另一个巨人的肩膀上。在沃顿商学院工作无疑是一种荣幸。因为我有幸和全世界的精英们在一起工作和学习。但无论多棒的工作，必然要渡过蜜月期。我无不例外地也被惯性蒙蔽了。在沃顿商学院工作了一年多，我开始对这里一切的优越工作环境和所有世界级的沃顿商学院教授都不以为然了。直到有一天，我问了沃顿商学院的一个教授（James Thompson）巨人故事的第二个问题："鉴于您已经站在巨人的肩膀上，这也就意味着您对这个巨人有一定的信赖。那么，您能告诉我这个巨人的个人性格特征是什么吗？"他没有直接回答我的问题。他说："杰克，我认为这个'巨人'不应该是单数。巨人应该是复数的。"在听到他的回答的那一刻，我忽然顿悟。深深吸了一口，并回答道："啊，那您就是我的巨人之一！"我们都会意地笑了。也在那个时候意识到这个著名的教授不是只站在一个巨人的肩膀上脱颖而出的。他无疑是一个站在许许多多巨人肩膀上的那个巨人。他不仅善于识别他生活中的诸多巨人们，更是一个卓越和谦卑的学者。通过这次对话，我真的自惭形秽。突然开始有意识地追寻过去曾经遇到的巨人们。无法抑制的感恩意识淹没了我！回首往事，所有那些我曾经遇到过的"巨人"们，是他们成就了今天的我。更深层次地反省，要感谢那些托起我的巨人的巨人们。以此类推，我们的国家、我们的地球，还有这个造就了我们生存空间的神秘宇宙都是值得我们感激的巨人。此刻，我油然地感受到一种无法言喻的万物归一的心情，情不自禁地遐想到了一个我无法回答的问题：到底谁是那个造就一切的终极巨人？我虽然没有答案，但是，我不得不说的是，这首我最钟爱的歌《You Raise Me Up》明确地表达了我现在的心情"……You raise me up so I can stand on mountains / You raise me up to walk on stormy seas / I am strong when I am on its shoulders / You raise me up to more than I can be(5-8)。（……你托起了我，所以我能站在群山之上 / 你托起了我，所以我可以经受惊涛骇浪 / 我是坚强的，当我站在你的肩膀上时 / 你托起了我，所以我才能超越自己。）"

应用练习

· 你可以回忆起上次你对一个人或一份工作不以为然的时候吗？是什么让你重新欣赏和珍惜那个人或那份工作？

· 你能列出一些你生活中遇到过的惯性、惰性事件吗？你是怎样意识到这惯性、惰性局面的，又是如何最终克服惰性的？

第 7 章

一个巨人的故事
（突破 – 转型）

去年，经我的一位美国朋友推荐，我和他在费城组织了一个海外中国留学生的领导力学习小组。我聚集了七八个感兴趣的学生（他们来自于费城的 Drexel、Temple、University of Pennsylvanian and Saint Joseph University 等大学）。由于我们学习小组内容的相关性和时间的约束，小组活动没能持续几个星期就搁浅了。但我没有把它当作一个失败。因为，毕竟那是一个很好的学习机会。从那时起，有很多问题一直纠缠着我："我怎样才能在很短的时间挖掘到人们潜意识里的价值观？我需要改善和提高自己哪些方面的教练辅导能力，才能成为对别人更具有影响力的辅助？我和那些学生之间的沟通桥梁又是什么？"这些问题让我急切地渴望知识和更好的解决办法。我试图回忆以前交流过的数百名学生，我发现他们大多数有一个共同的主题。他们都或多或少地对权威人物或社会有一定的失望（很像我当年年轻的时候），忽然，我感到了与他们之间的共鸣。也就是在几个月前，我开始问一系列的互动问题，以测试或检测被问方的个人价值。我相信大多数人看重和在乎的内在价值会决定他们的愿景、目标和行为模式。一开始，我只是想知道谁是那些学生佩服的巨人。但是我越问就越好奇他们的回答。然后，教授Janet Greco故事课让我找到了写这本书的最好借口。于是，我就以"追寻巨人"为主旨写此书。我相信此书会帮助任何一个读者发觉自己潜意识里的一些珍贵的东西。

书看到这里，你一定对巨人的比喻非常熟悉了。"巨人"绝对不是一个新的概念。孔子曾经说过："三人行，必有我师焉。"在中国古代，老师被认为是父亲，"一日为师，终生为父"；鲁班学艺的导师或教练等等。孔子描述"巨人"往往在我们中间，用谦卑的心，我们就可以识别身边的智者并从他们那里汲取到智慧。就在最近,我在和我的一个同事分享巨人理念时,当她听到了我的想法后,她立即推荐给我读 Malcolm Gladwell 的书《Outlier》(Gladwell, 2008)。在这本书里,作者雄辩地介绍了很多成功人士,比如:比尔·盖茨、史蒂夫·乔布斯,还有犹太社区等成功人士或人群的诸多例子。他们的成功或成就主要是借助于他们的家庭背景、文化的遗产、顶级的养育,还有他们生活中的导师或教练等必要因素。我强烈建议大家一定要拜读此书。因为书中的 Outlier 及局外人的概念和我书中提到的巨人的概念几乎是完全一致的。

显然,巨人可以是领导者,是导师,或是帮助我们提升的教练。巨人也是这个世界的积极因素(如文化的遗产、有利的环境)不停地推动着世界往前发展。很多时候,这些巨人在不同时期以榜样、导师、教练或有利的环境或文化传统等因素出现在我们的生活中。最终,我们也在生命的号召下渐渐地成长为下一代巨人,只有这样我们才能够提供自己的肩膀让年轻人或者下一代站在上面。我们注定要成长并引领一个时代,并迟早接受做巨人的使命。

为了更好地把巨人这个实用的概念结合我们每个人的实际生活,并有效地把这个概念运用到我们的日常生活中,我想先和大家一起简要地回顾一下巨人的故事问卷。之后,这个问卷将完全属于每一个人,这意味着你可以自由地对任何人进行这项问卷调查。我相信,只要你愿意花 5-10 分钟的时间与你周围的人交互式地进行这项调查问卷,每次都会有不一样的神奇事情发生。如果你有任何怀疑的话,不妨试试看。

1. 假设你站到了一个巨人的肩膀上,你会看到什么,又会感觉到什么?

愿景练习——这是一个奇迹类问题,目的是让我们锻炼愿景能力。这个问题帮助我们从更高、更开阔的视野看世界,同时意识到我们自己的渺小。我鼓励你对尽可能多的人问这个问题。我向你保证,他们的回答会给你很多启示。这个问题不但能锻炼人们的想象力和愿景，最重要的是，它会大大提高人们描述和沟通愿景的能力。

2. 既然你站到了这个巨人的肩膀上，也就意味着你和巨人之间有一定程度的信任。所以，请你列出 3 到 5 个形容词来表达这个巨人的个人性格特点是什么？

发现和探索人们潜意识里的价值——这个问题旨在逆转巨人的视角，并聚焦到巨人的内心世界。它帮助我们看到参与者自身的内在价值，这些价值也正是投射到巨人身上的个人性格特征。这些特征是人们的原始价值或起始价值。这个问题的目的是为了让我们意识到自身的内在价值，并最终鼓励人们发展和扩大自己的价值库。

James Kouzes 和 Barry Posner 花了超过 25 年的时间对领导力做了研究。在对超过 75000 人的问卷调查取得的结果中,他们总结出领导力的特性是"诚信、前瞻性、启发性、智能化、公正、心胸宽广、支持、直观、可靠、合作、坚定、富有想象力、有事业心、勇敢、有爱心、成熟、忠诚、自我控制、独立"等等(引自 Wildflower & Brennan, 2011, 第 149 页)。有趣的是,在我收集的有限的《巨人的问卷》的答案里,我发现了这些答案与 James Kouzes 和 Barry Posner 调查结果有惊人的相似之处。下面的图表是我收集巨人故事问卷中问题 2 的一些答案,供大家参考。

巨人的个人性格特点是什么？

年龄	性别	职业	问题2答案
20+	女	学生	善良、执着和坚贞
20+	女	学生	领导、爱心、智慧和幽默
20+	女	学生	成熟、严谨、纪律、有爱心
20+	女	学生	贴心、可靠的
30+	女	翻译	勇敢的、可靠的、有爱心
30+	女	律师	可靠的、安全的和有逻辑的
40+	女	高管培训	乐于助人、良好的追随者、支持
40+	女	高管培训	善良、温柔、富有同情心、坚强、谦虚
40+	女	律师	可靠、高纬度、富有远见
40+	女	学生	聪明、善良、开明、经验丰富和神圣
50+	女	退休	功能强大、值得信赖、可靠、问责
40+	女	高管培训	沉稳、谦虚、坚强、无私、体贴
50+	女	护士	温柔、耐心、力量
40+	女	总裁助理	可靠的、有耐心、有见地
40+	女	客户经理	强度、保护、值得信赖
30+	女	老师	爱心、稳定、强大的、诚实的、有知识
40+	女	研究员	辉煌、牢固、可靠、接地气
30+	男	经理	支持、技术过硬、平衡和有良心

续表

年龄	性别	职业	问题2答案
20+	男	学生	可靠的、值得信赖和强大
20+	男	学生	知识渊博、聪明、负责、宽容、忠诚、善良
20+	男	学生	稳定、值得信赖、陪伴
30+	男	高管培训	有经验、有爱心、透视
30+	男	咖啡商	爱心、带动、忠诚
40+	男	教授	温柔、大方、乐于助人
40+	男	教授	功能强大、宽容、高度、平衡
20+	男	人事经理	善良、大方、幽默
20+	男	高管培训	强大的、支持的、比生命更大、安慰
30+	男	高管培训	积极、授权、纯粹和关怀
30+	男	制药	自信、无摇摆不定、骄傲的、能耐

老实说，当我第一次把以上的答案再归纳在这个电子表格里时，我非常惊讶于所有这些参与者的巨人的个人性格特征。他们的性格特征都是那么美好、令人振奋并具有积极向上的意义。在我惊喜的同时，我又反问自己："我此前为什么没有看到所有周围人身上的美丽心

灵？"如果我没有问他们这个关于巨人的性格特征的问题，可能永远被蒙在鼓里，并可能对他们的美丽的内在价值一无所知。这也似乎是为什么连自己在内的许许多多在普通岗位上努力拼搏的我们总渴望被领导重视，但是却常常感到不被重用，或被忽视的冷落感。因为，现实生活中的确是千里马常有，伯乐难求。有一点我特别清楚，那就是几乎每一位参与巨人故事问卷的人都拥有一些 James Kouzes 和 Barry Posner 上述研究中提到的领导特质。此刻，我有个疑问：如果千里马始终没有遇到伯乐的赏识，是否千里马就慢慢地平庸下去了呢？试想一下，如果所有的参与巨人故事问卷调查的人有意识地并主动地选择把自己的情感、思想、行为和自己的"巨人"特质调节一致的话，他们又将成就怎样的领导人？他们又将怎样给这个世界带来积极的影响？说到这里，我突然有一个大胆的设想，如果我们每个中国人都找到了自己的内在巨人，并明确地意识到了自己巨人的性格特征（价值）。那么，中国这个东方巨人身上的正能量，整个世界都会有目共睹。

3. 如果你站在肩膀上的那个巨人跌倒了,你会怎么做?

变更管理——人们往往在很大程度上会受到其周围环境的影响。这个问题是为了帮助人们意识到自己对突发事件的最初反应。希望这个问题能帮助参与者意识到自己对突发事件和当时环境的即时反映,能设想出多个建设性的应对方案,并最终选择最佳方案。"巨人跌倒了"可以联想到我们生活中的巨人:家长、导师、权力机构或社会等等。如果令我们失望了,我们会怎么做?这个问题我没有例举任何例子是因为我想让大家自己去获取参与者的回答,并自己体会这些答案的奥妙。

对于问题3,我希望大家要谨慎地提问,因为它可能会触及到参与者的内心世界的一些敏感点。强烈建议大家对这个问题的答案要保密,并尽量不要和参与者以外的人分享他们的答案。

4. 如果你看到另一个巨人行走在你现在所站在肩上的巨人附近,而且另外一个巨人要比你现在的这个巨人高大得多,在此种境况下,你会做些什么?

驱动力和动机——野心和上进心。我问这个问题本来是想检测参与者是否愿意走出自己的舒适区并挑战自己的成长。但事实

上,这个问题的许多答案都非常耐人寻味。例如:

- "哇,另一个巨人。虽然,他比我的巨人还要高大,但不清楚是否可以信任另一个巨人。我还是比较放心自己心中的巨人。"

- "我会从我的巨人肩上跳下,然后爬到另一个巨人的肩上。"(我后面紧接着又问了一个问题:如果另一个巨人不让你上他的肩膀你怎么办?)

- "我会请我的巨人和另一个巨人交流。当他们成为朋友时,我就可以请求我的巨人借另一个巨人的胳膊,这样我可以顺着另一个巨人的胳膊爬到对方的肩膀上去。"(这个答案显示了参与者不凡的社交情商。这是我十岁的儿子乐星给出的答案。)

- "我会问我的巨人,你认识另一个巨人吗?如果他觉得另外一个巨人很可靠,我可以翻越到另一个巨人的肩膀上。"

这只是几个回答的例子,我猜你可能已经开始解读这些答案的字里行间的涵义了。提问是一个非常复杂、有趣的形式。它不仅可以帮助我们理解别人,还能进一步加深我们对自我和他人存在的意识感。我相信学习是一个双向的过程。每次进行这项调查问卷,我都会发现自己在不停地学习和成长。

5. 如果你站在巨人的肩膀上太长时间后，会发生什么？

处理惯性、惰性——这个问题的目的是为了使我们对自己的个人和工作、生活中的惰性产生意识，并克服它。正如我们所知道的，惯性、惰性会时常发生在每个人、每对夫妇、每个婚姻或是每个组织里面的一个普遍存在的现象。为了让大家更好地了解"惯性、惰性"，我们最好先看看这个词的含义：

Inertia 惯性、惰性

Noun 名词

① the state of being inert; disinclination to move or act 是惰性的状态；懒的移动或行为

②（physics）（物理）

a. the tendency of a body to preserve its state of rest or uniform motion unless acted upon by an external force. 物体保持静止或匀速运动的趋势或状态，除非采取外力改变行动趋势。

b. an analogous property of other physical quantities

that resist change: thermal inertia (《Collins English Dictionary》, 2012). 热惯量(《柯林斯英语词典》, 2012)等物理量抵制变革的类似性。

不难想象，当我们呆在巨人的肩膀上很长一段时间后，我们将不可避免地陷入惯性或惰性阶段。从而，我们开始对自己选择的特权和自身优越的环境感到理所当然，或是不以为然。基本上，我们开始对优越的现实变得麻木而失去了感恩上进的心。我在沃顿商学院的惰性经历就完全证明了这一现象。同样，我们在人际关系、婚姻和工作中都很容易在蜜月期后步入惯性或惰性区间。如何应对和处理这个惯性、惰性区间对我们个人的成长和所属群体以及组织机构的存亡和升华都具有至关重要的意义。

当你完全理解这些巨人问题背后的理论后，我相信你完全可以在你周围的人群中进行这个巨人故事问卷调查。所以，**go rock-n-roll**（这个舞台是你的了，尽情发挥吧）。我强烈地建议你一定要给至少 10 个以上的人做这个巨人故事问卷调查。其中的奥妙和神奇只要你做完十几个人的问卷调查后,你自然可以体会到。

应用练习

下面的问题是在五个巨人故事的核心问题之外的一些附加问题。当你对基本问题应用自如后,我建议你大胆尝试下面的问题。我相信这些问题会对接受《巨人的故事问卷调查》的人带来更深刻的启发和引导。

• 当你的巨人摔倒后,除了你所说的那个行为,你还能设想出其他几个选择方案吗?

• 认识和接触到另一个(或其他)巨人的最佳方法是什么?

• 我们自己怎样才能成为一个值得别人依靠的巨人?

• 在你的家人面前(如果你已成家),你会为自己的家人们展现出的巨人性格特征是什么?

- 你和你理想的巨人之间的差距是什么？你打算如何弥补这之间的差距？
- 我们如何实现价值观、信念和行动之间的一致？
- 你打算如何迈出你的"巨人步"？
- 怎样才能更有效地发现我们身边的巨人？
- 巨人与自我优越感（ego）之间的区别是什么？你可以回忆一下当巨人被 ego 蒙蔽的时刻或例子吗？
- 怎样才能扩大和增加更多优良的巨人的性格特征？
- 回忆一下你曾经遇见过（过去和现在）的巨人（如果时间允许的话），分享一下你遇见的每个巨人。

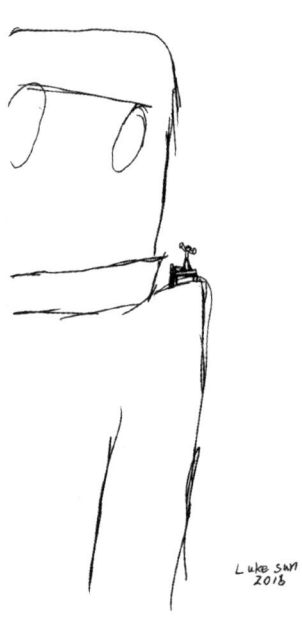

第 8 章

生生不息的巨人传奇
（实际应用 – 唯美和功效）

根据过去几个月里我做的巨人故事问卷调查，我归纳了以下观察结果：

- 巨人和每个参与者之间都有下列有趣的关系：相互信任、相互依存、相互双边关系。有些参与者偶尔感到对巨人的不信任和失望感。这本书中我没有提及巨人的情感需要。其实，巨人也是人，他们也需要情感支持，他们也会像我们每个人一样需要被信任、被尊重。他们有时也会感到受伤和失望。

- 多数年轻一点的参与者对巨人的理念有点陌生。而年长的参与者，对于巨人的理念有极强的共鸣。

- 巨人的性格特征是无形的，但是如果我们给予定义和认可，并使大家对之产生强烈明显的意识，这将极大地影响我们每一个人的生活和工作。

- 意识到巨人的性格特点会对我们自己的行为、动机和生活目的都带来不可估量的启示。
- 意识到我们彼此之间的巨人将极大地帮助个人、群体和组织，建立更具协作性和互爱的社区。

这些观察结果只是在我收集的有限软数据里一点儿管窥之见。但是，我希望邀请你们每个人进行这个巨人故事问卷调查。我相信你们会将亲自体会这个强大工具，以及这个工具为你和你周围的人带来难以想象的积极和深远影响。我敢打赌，你将更深刻地认识到巨人是如何与我们每个人、团体、组织，或整个社会安定和进步息息相关的。

在进行巨人故事问卷调查中，我想推荐大家继续专注于以下三个方面的研究：

- 内心的巨人——无意识的自我和运作中自我之间的关系

- 交际中的巨人 ——自我和导师、教练、楷模之间的关系
- 组、团队、企业巨人 —— 自我和组织、集体或企业机构之间的关系

下面的内容中,我把三个方面的研究分别简要地解析。希望我的推理会激励你采取必要行动。从而,使你的工作和生活变得更美好。

内心的巨人

自古以来,意识自我都一直被认为是一种智慧。2500多年前,老子在《道德经》里提到:"知人者智,自知者明。"同样,Daniel Coleman 在他的书《Focus》中非常犀利地论证了对自我、他人和外界意识是一个人出类拔萃的基本(Coleman,2011)。我们常听人说"我们做人要有自知之明",也是同样的道理。自知之明说起来容易,做起来难。所以,我希望通过这个巨人故事问卷的实际练习,你我都会更加清晰地意识自我。此外,

也希望我们对自己内心的巨人,还有我们身边的巨人,以及外界的组织结构巨人都有一个更深层的认识。对我来说,内心巨人的成长和成熟是将来升华和成为交际巨人和外界组织机构巨人的基础。

人际交往中的巨人

有一天,我让一对情侣做了巨人故事问卷调查。在他们的巨人性格特征的答案里,无意中发现了他们不断争吵的根源。请先看一下他们各自的答案:

巨人的个人性格特点是什么?

	Bob(化名)	Teresa(化名)
巨人的个人性格特点	知性、睿智、守信、负责、宽容、忠诚、善良	成熟、严谨、有教管、关怀

当我问道 Bob 有关巨人的个人性格特点时,他马上问到我所说的那个巨人是不是指他的父亲?顺着这个思

路，他轻松地回答了巨人的性格特征问题。从他的答案里不难看出他的父亲是一个很棒的导师，而且他们之间有很深的信任度。按照 Bob 自己的话，他的父亲通过真实亲历沟通分享了他的人生智慧。Bob 自己具有和他同龄人相比之下少有的安全感和自信。他不但有极强的责任心，又不缺乏冒险精神。Teresa 和 Bob 交往至今有近两年时间了，已经达到了谈婚论嫁的程度。Teresa 是一个非常有爱心、善良的人。她对 Bob 也疼爱有加。然而，根据她的巨人性格特征（我在问巨人故事问卷时 Bob 和 Teresa 都彼此不在场，这对这个问卷调查很重要），她的巨人似乎缺乏安全感。我不清楚在她年轻时发生过什么事导致以上结果。但这个结果让 Bob 和我都突然明白为什么 Teresa 总抱怨 Bob 不够成熟，而且她有一种强烈的倾向去控制 Bob 社交生活（严格、教管的行为）。从此看来，Teresa 的巨人特性和 Bob 的巨人守信、宽容和忠诚特性恰恰相反。我可以推断 Teresa 需要克服自己内心某种恐惧才能对 Bob 充分地信任和接纳。同样，Bob 也要学会去接纳 Teresa 和她不同的原始价值趋向。

从他们身上收集到上述信息之后，我们都坐在客厅里探讨了他们彼此的巨人。首先，我把 Teresa 和 Bob 的巨人的性格特征都呈现在他们面前。当他们俩得知对方的巨人的性格特征后都有种说不出来的顿悟。探讨到最后我说道："如果你们两个想踏踏实实地建立一个家庭的话，你们俩就必须要重新建立或创建一个新的巨人。这个巨人将会结合或创新出你们彼此都同意的价值观。这样的话，你们两个巨人的性格特征将会继续扩充，而这会极大地稳定和充实你们的情感关系以及深层次的精神交流。我相信基于现在你们对彼此巨人的认识或意识，你们俩的未来一定会很美好。"

人际交往中，一对夫妇或一个家庭，他们都需要一个巨人或领导负责为这个小团体设想出一个共同愿景，以及建立和创造这个小团体共同的价值观。我相信，只有这样，夫妻或家庭团体才很可能活得更充实、更有意义。人际交往中的巨人对我的启示就是你我都可能成为别人的巨人，所以从我做起必然是硬道理。也就是说，我们在清楚地意识到我们内心的巨人后,我们要让这个巨人在现实生活中经历种种考验和磨练并在我们的呵护下成长为别人可以信赖和依靠的巨人。

公司情景中的巨人(组织巨人)

当谈及公司情景中的巨人时,我情不自禁地想到了 GE 的前 CEO:Jack Welsh。在他的书《Straight from the Gut》的开场白中 Jack 写道:

> "媒体曾给过我很多外号,那些外号完全不能体现我是谁,而是更贴切地说明了我们通用电气公司所经历的各个阶段。事实是,从外到里,我从来都没有怎么改变过。我仍旧是我母亲抚养大的那个在马萨诸塞州塞勒姆出生的男孩(第 XV 页)。"

他的话清楚地告诉我们,作为一个成年人,Jack 仍坚定不移地保留他的妈妈从小言传身教给他的价值观

(天堂的财富)。他在书中这样叙述:

"当我1981开始在通用电气的旅程那年,我第一次站在纽约华尔街的皮埃尔酒店前面对分析师说,我想让GE成为'全球最有竞争力的企业'。我的目标是把小公司精神转化到大公司身上,并把一个老牌的工业通信公司建立成一个比GE小1/50的公司的精神更加高昂的公司,经营更加灵活的公司(第XV页)"。

从杰克·威尔士的陈述中我们不难看出,他已经成功地把他自己的精神和价值观成功地灌输到了他之前任职的小公司身上。而且,他当年也满怀信念地要把他所谓自己身上的小公司的精神带到通用电气这个大公司身上。很显然,杰克·威尔士的逻辑是试图从根本上把他的精神和价值观植入到小公司之后再复制到一个更大的GE公司身上。

这个公司或组织结构下的巨人对我的启示是,一个可持续发展的公司里一定要有一个可持续性成功的领导者或巨人引航，而这个领导者或者巨人和我们每个人一样都是从一个小孩子慢慢成长起来的。如果你注意到 Jack Welsh 提到了他的母亲。我想你一定也意识 Jack Welsh 的母亲就是他生命中的巨人之一。

从上面的讨论中，我想读者朋友对巨人故事问卷的实用价值有了一定了解。这个巨人理念不但对我们每个人，对每个团体和组织或公司都有不可估量的实际意义。我真诚地希望这个巨人的故事问卷可以给我们每个人的学习或工作、生活都带来可以利用的价值。

这本书是以巨人的故事问卷为开始，最后，我想以另一个巨人的故事问卷结束本书：

巨人故事问卷 2

1. 想象一下，如果你的伴侣、家庭、组织或公司把你当作他们可以依靠的巨人时，他们站到你的肩膀上，能看到什么，又能感觉到什么？

2. 如果让你的伴侣、家庭、组织或公司用 3 到 5 个形容词来形容你的个人性格特征，他们会说出或写出哪些形容词？

3. 如果你的伴侣、家庭、组织或公司对你感到失望，你会如何选择与他们沟通或处理他们的情绪和人际关系？

4. 如果你的伴侣、家庭、组织或公司不满足于你的领导或引导，你会怎么做？

5. 当你的伴侣、家庭、组织或公司被惯性或惰性状态卡住时，你会怎么做？

附录 1

期　望

编写《巨人的故事问卷》目的，在于帮助大家能更明确地意识到我们每个人的内在价值，并在此基础上追述自我价值的沉淀，再以愿景的方式预见我们成为巨人后的自己。最终，通过巨人的故事理念把我们的过去、现在和未来的巨人自己，链接到当下，从而做出具有战略意义的选择，并不断地学习和完善我们内心的巨人。

我衷心地希望每个读者都能开始意识到自己的巨人性格特征（价值）。因为，这些价值是你我生命赖以依靠的精神支柱，看清我们自己内在的巨人性格特征，并选择将这些特征作为你我人生故事的主旋律，以身作则地书写一个属于我们自己的唯美巨人故事。你我都是这个故事的主角，我们每一个选择都决定了这个故事的方向。选择正能量，并让我们的巨人价值库不断累积通往天堂的财富。当我们能够成就自己巨人传奇的时候，正能量的价值自然会传承下去。藉此，一群集体巨人，譬如：一个家庭、一个社区、一个组织、一个国家或者一个世界——自然而然也将得以成就。

我们每一个人都有自己独一无二的巨人故事。我相信：这个巨人故事问卷不但可以成为我们追寻巨人旅程的开始，也一定会帮助我们每一个人开启各自的领导力探索和发掘的旅程。同时，希望在巨人探索之旅过程中，下面几个问题可以不时地提醒到你：

怎样像巨人一样思考？

怎样做到像巨人一样的做事风范？

怎样设想出巨人的愿景？

怎样和巨人们成为朋友？

作为巨人，我们如何回馈社会？

寻求和成就内心的巨人，是我们每个人一生的学习和成长旅程。无论你现在旅程到哪个阶段，请一定继续谦卑地去结识更多的巨人，大胆地栽培新生代巨人，并由衷地期望每一个属于我们自己的巨人传奇生生不息地传承下去。

附录2

我的中国梦

曾经,有不少人问我:"你的美国梦是什么?"我对他们的回答是:"我没有美国梦,但我倒是有中国梦!"我梦想中国这个精神上的巨人从东方崛起。我梦想中国有最幸福的家庭、最善良的老百姓、最美丽的河川。我梦想着中国能把五千年文化里最灿烂的正能量传承下去。其实,我们每一个人都是中国这个巨人身上的一个细胞。当你我都成就了自己个体巨人的时候,也就自然而然地成就了我们"中国"这个精神上的巨人。

我八岁的小儿子偶尔听到我问其他人关于巨人的故事问卷时，他把他自己的巨人画了出来。你可以看到他还不会拼写"Truth"、"Sword"、"Unity"、"courage"和

参考书目

"Building Companies to Last." *Jim Collins*. Retrieved December 10, 2014, from Jim Collins' website: http://www.jimcollins.com/article_topics/articles/building-companies.html

Gladwell, Malcolm. *Outliers: The Story of Success*. First Edition. Hachette Book Group, NY, 2008. Print.

Goleman, Daniel. *Focus: The Hidden Driver of Excellence – First Edition*. HarperCollins Publishers, NY, 2013. Print.

inertia. (n.d.). *Collins English Dictionary - Complete & Unabridged 10th Edition*. Retrieved December 09, 2014, from Dictionary.com website: http://www.dictionary.reference.com/browse/inertia

Johnson, Ian. "Train Wreck in China Heightens

Unease on Safety Standards." Retrieved December 10, 2014, from The New York Times website: http://www.nytimes.com/2011/07/25/world/asia/25train.html?_r=2&

Kegan, Robert, and Lahey, Lisa L. *Immunity to Change: How to Overcome It and Unlock Potential in Yourself and Your Organization.* Boston, MA: Harvard Business, 2009. Print.

Kouzes, James M., and Posner, Barry Z.. *The Leadership Challenge.* San Francisco, CA: Jossey-Bass, 2007. Print.

"*Lao Tzu.*" Retrieved December 10, 2014, from Know Thyself website: http://thyselfknow.com/lao-tsu/

"TechTalks #2 Inspirational Video: First Follower Leadership Lessons from Dancing Guy." Retrieved December 10, 2014, from YouTube Website: https://www.youtube.com/watch?v=1euTzm4-pMU

Welch, Jack, and Byrne, John A.. *Jack: Straight from the Gut.* New York: Warner, 2001. Print.

Wildflower, Leni, and Brennan, Diane. *The Handbook of Knowledge-based Coaching: From Theory to Practice*. San Francisco, CA: Jossey-Bass, 2011. Print. (p.149.).

"TechTalks #2 Inspirational Video: First Follower Leadership Lessons from Dancing Guy." Retrieved December 10, 2014, from YouTube Website: https://www.youtube.com/watch?v=1euTzm4-pMU

Welch, Jack, and Byrne, John A.. *Jack: Straight from the Gut.* New York: Warner, 2001. Print.

Wikipedia. N.p., n.d. Web. 24 Apr. 2015. Retrieved from http://en.wikipedia.org/wiki/Giant_(mythology)

Wildflower, Leni, and Brennan, Diane. *The Handbook of Knowledge-based Coaching: From Theory to Practice.* San Francisco, CA: Jossey-Bass, 2011. Print (p.149).

Unabridged Tenth Edition. Retrieved December 9, 2014, from Dictionary.com website: http://www.dictionary. reference.com/browse/inertia.

Johnson, Ian. "Train Wreck in China Heightens Unease on Safety Standards." Retrieved December 10, 2014, from The New York Times website: http://www. nytimes.com/2011/07/25/world/asia/25train.html?_r=2&.

Kegan, Robert, and Lahey, Lisa L. *Immunity to Change: How to Overcome It and Unlock Potential in Yourself and Your Organization.* Boston, MA: Harvard Business, 2009. Print.

Kouzes, James M., and Posner, Barry Z.. *The Leadership Challenge.* San Francisco, CA: Jossey-Bass, 2007. Print.

"Lao Tzu." Retrieved December 10, 2014, from Know Thyself website: http://thyselfknow.com/lao-tsu/

Merriam-Webster. N.p., n.d. Web. 24 Apr. 2015. Retrieved from http://www.merriam-webster.com/ dictionary/value

Bibliography

Baiki. N.p., n.d. Web. 24 Apr. 2015. Retrieved from http://www.baike.com/wiki

"Building Companies to Last." Jim Collins. Retrieved December 10, 2014, from Jim Collins' website: http://www.jimcollins.com/article_topics/articles/building-companies.html.

Gladwell, Malcolm. *Outliers: The Story of Success*. First Edition. Hachette Book Group, NY, 2008. Print.

Goleman, Daniel. *Focus: The Hidden Driver of Excellence* – First Edition. HarperCollins Publishers, NY, 2013. Print.

Grant, Adam M. *Give and Take: A Revolutionary Approach to Success*. New York, NY: Viking, 2013. Print.

Inertia. (n.d.). *Collins English Dictionary - Complete &*

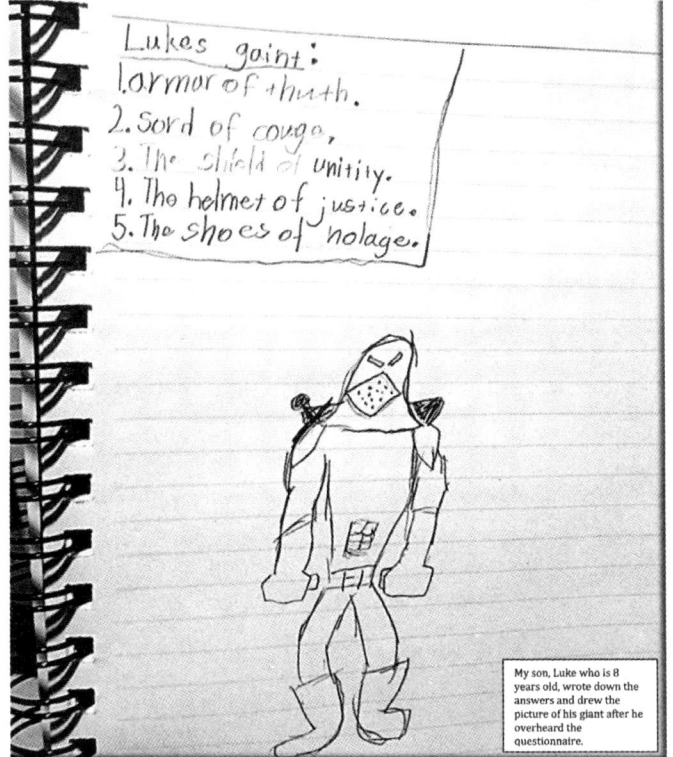

Lukes gaint:
1. armor of thuth.
2. Sord of couge,
3. The shield of unitity.
4. The helmet of justice.
5. The shoes of nolage.

My son, Luke who is 8 years old, wrote down the answers and drew the picture of his giant after he overheard the questionnatre.

Appendix 85

be honest that it is the ultimate Giant I believe in who led me through all the doubts and fears in me. I was once lost until my German friend lit up my dark world and the Giant was there. When I was discouraged and powerless, the Giant was there. It is because this Giant who never falls and always believes in me. Some call this Giant love; some call this Giant life; some call this Giant peace; some say this Giant is omnipotent; some say this Giant is eternal.... Well, I see this Giant as the Giant behind all the giants. Not sure what this Giant looks like, but I do know the values we hold dear to my heart are the reflection of this Giant. The most important thing is that no one can rob these values away from us and no one can steal them from us, unless we give them up.

Appendix

I still can't believe at this moment that I am writing a book and this is the book I am writing now. This giant idea came to me serendipitously, and somehow, I listened with my heart and chose to let it grow and expand into a tree. It may sound too good to be true. It is, however, happening. The real story is that there were so many doubts in myself from the time I conceived the idea to write this book. I feared my professor wouldn't like my writing, especially my grammar mistakes. I feared people would laugh about my giant idea or concept (it did happen many times). I feared I would quit during the middle (Thanks to Steve Shepard's encouraging words, I didn't). I feared no one would appreciate this book (still not a hundred percent sure). What has motivated me to this moment? I have to

on your shoulders what they would see and how they would feel?

2. If given 3-5 adjectives from your partner/family/organization, what would they say about your personal characteristics?

3. If you made your partner/family/organization disappointed or let them down, how would you choose to communicate with them?

4. If your partner/family/organization was not content with you, what would you do?

5. What would you do if your partner/family/organization was stuck in inertia?

the organization's core values.

It is probable to draw the conclusion that an ideal organization not only needs a leader/giant who is a visionary, but also one who commands this giant to embody his vision, values and actions authentically to the organization's structure, system, strategy and goals while building a culture that aligns with everything else.

From the above discussion, it is apparent that the Giant Story Questionnaire has applicable and practical implications toward individuals, groups and organizations. I sincerely hope this Giant Story Questionnaire can be greatly harnessed to contribute to people's personal growth as well as an organizational effectiveness and boost human potential in the work place as well as other areas of our lives.

As I started with the Giant Story Questionnaire, I want to end this book with another Giant Story Questionnaire:

1. Imagine your partner/family/organization is standing

intent to replicate to a much bigger company.

Complimentarily, the author of *Built to Last* James Collins (1995) argued that building a visionary company demands a visionary leader who is a clock builder and an architect, but not a time teller. The visionary company not only embraced the genius of the "and" adversary and thrive above ambiguity, but also preserves the core ideology, while stimulating progressive movement ahead. Most of all, the visionary company has to consistently align with the organization's core values and with it's goal, culture, structure and progress. In his book he makes a comparison of visionary companies to a group of good companies. He found that charismatic leadership was not the differentiating variable. The Visionary Companies survive, visionary leaders come and go and so do visionary products. His findings really help us to understand that building a visionary company demands more than just a visionary leader but someone who can align everything to

that as an adult he still held steadfast the values his mom or parents breathed in him when he was a little boy. As he continues,

"When I started on this journey in 1981, standing before Wall Street analysts for the first time at New York's Pierre Hotel, I said I wanted GE to become 'the most competitive enterprise on earth'. My objective was to put small company spirit in a big-company body, to build an organization out of an old-line industrial company that would be more high-spirited, more adaptable, and more agile than companies that are one fiftieth our size." (p. xv).

Basically, Jack Welsh indicated he had successfully embodied his spirit or values to a small company-body. And then he intended to put the small company spirit in a big-company body. It's not hard for us to see the pattern behind Jack Welsh's logic. He was fundamentally trying to implant his spirit and values to his company, and then

ahead of you."

Interpersonally, in a couple or in a family, it demands a giant/leader who is responsible to envision a shared future and create shared values. I believe only by doing this, the couple or family will be very likely to live a more fulfilled life.

Giant in the corporate setting (Outer Giant)

When we talk about the giants in the corporate setting I cannot help myself from referring to GE during the time of Jack Welsh as its CEO. In the prologue of Jack's book: *Jack: Straight from the Gut* (2001) he stated that after the medium called him many names: "Those characterizations said less about me and a lot about the phases our company went through. Truth is, down deep, I've never really changed much from the boy my mother raised in Salem, Massachusetts." (p. xv). This comment clearly showed us

exactly sure what happened to her when she was younger. Based on the characteristics of her giant it appeared to me that she demands Bob to grow more in maturity, and she has a strong tendency to control (strict, disciplined), which is very opposite from Bob's trustworthy, forgiving and loyal characteristics. I am assuming there is something that stopped her to fully trust and accept Bob as who he is.

After I gathered the above information from them respectively, I had the two of them sit in their living room, and presented their giants' personal characteristics in front of them. It was more like an Aha moment for both them. I commented at the end of our discussion, "If both of you want to start a family together soon you have to build or create a new giant who will incorporate both of the values you would like to keep from your old giants and continue to expand this giant's characteristics repertoire to secure and enrich your relationship. I am confident with this awareness you two are going to have a beautiful future

	Bob (not real name)	Teresa (No real name)
Giant Characteristics	Knowledgeable, wise, trustworthy, responsible, forgiving, loyal and kind	Mature, strict, disciplined and caring

When I asked Bob about the personal characteristics of the giant he immediately asked me if I was referring his father as the giant. Nevertheless, he answered the question with ease. His answer indicated that he is well mentored or coached by his father. His father and he have a great trusting relationship. He shared all his wisdom through authentic communication according to Bob's own words. He feels secure and confident with himself. He is not afraid of taking risk while being responsible. Teresa and Bob have dated for almost two years and have reached the point of talking about wedding days. Teresa is a very caring and kind person who loves Bob dearly. However, she is insecure with their relationship somehow. I was not

since ancient times. Lao Tzu said in Tao Te Ching more than 2,500 years ago, "Knowing others is intelligence; knowing yourself is true wisdom." (Lao Tzu, n.d.). Likewise, Daniel Coleman illustrated in his book Focus that self, other and outer awareness is the foundation for excellence (Coleman, 2011). I hope this Giant Story Questionnaire will bring unclouded awareness not only toward our own inner giant but also those who are around us. To me, the inner (intra-personal) giant is the foundation of the interpersonal and outer world giant.

Interpersonal Giant

One day, I was asking a couple I knew about their giant's personal characteristics. I accidentally found out the root of their constant quarrelling. Please see their answers for the questions listed as below:

While conducting the Giant Story Questionnaire, there are three parts of the continuing research that I would recommend to us to focus on:

- Intra-personal giant — relationship between unconscious self and functioning self
- Interpersonal giants — relationship between self and mentor/coach/role models
- Group/team/corporate giants — relationship between self and collective giants in a big congregation or organization

Respectively, I have put those three areas into perspective. I hope my reasoning will probe you to take prompt action toward making your work and your life a better place.

Intra-personal giant

Being aware of ourselves has been considered wisdom

the participants with increased age, the giant's values become more concrete or aligned to their consciousness.

• Giants are invisible, but if defined and recognized they will greatly impact one's personal life and work life.

• The giant's personal characteristics give us great insight for our own behaviors, motivations and purpose of our life.

• Being aware of each other's giant will help the individual, pair, group and organization be more community driven and more collaborative in nature.

These findings were just a glimpse of my observations based on limited soft data I gathered. I hope, however, by inviting each of you to conduct this Giant Story Questionnaire you will personally harness some unimaginable findings from this powerful tool. I'd bet you will have a much more profound understanding of the practical application of how crucial each GIANT is to an individual, a group and organization or the whole society.

Chapter 8:

The Virtuous Circle of Life
(Application - Aesthetic and instrumental)

Based on the Giant Story Questionnaires I have done for the past few months I have found the following observations:

• There is an interesting relationship between the giant and each participant: mutual trusting, interdependence, bilateral. Occasionally, broken, mistrust and hurt by the giants.

• Younger people are sometimes less conscious about the values projected by their giants (among the participants I found only few well-mentored young people who are well-aware of the concept of the Giants). Among

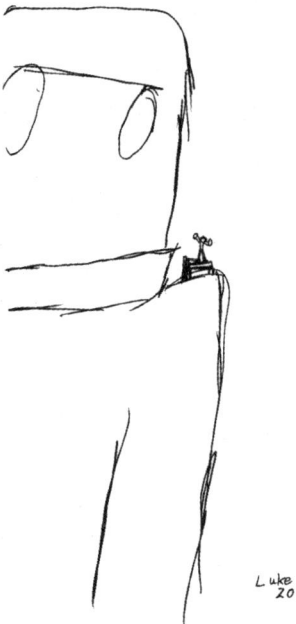

Luke Shn
2018

Chapter 7: The Giant's Story Questionnaire (Breakaway - Transformation)

beliefs and actions?

• What is your next "giant" step?

• How can we spot the giant around us?

• What would be the difference between the giant and the ego? Can you recall a moment that the ego took over the giant in your life?

• How can we expand the repertoire of the giant's personal characteristics?

• Who are the giants you have met in your life (past and present)? (If time permits) Would you mind sharing some of the defining moments you had with the giants you met in your life?

Application Exercise

The following questions are additional questions for the five core giant questions. As you become more comfortable with the basic questions, I believe the following questions can be of great assistance to you for coaching or mentoring toward the people you are conducting the Giant Story Questionnaire with.

• What other options would you have after the giant fell?

• What would be the best way to reach out to another giant?

• What would it take to be a giant for others to depend on?

• What are the personal characteristics you would like to embody as a leader in your family?

• What are the possible gaps between you and your perceived giant? How do you mend the gap?

• How do we achieve the alignment between our values/

uniform motion unless acted upon by an external force

b. an analogous property of other physical quantities that resist change: thermal inertia (Collins English Dictionary, 2012).

It is not hard to imagine that when we stand on the shoulder of giant for a long time, we will fall into a stage of inertia, where we take for granted the privilege we have. Basically, we start becoming unconscious of the blessings we have and unappreciative of the environment and life we are in. My Wharton inertia experience has perfectly demonstrated this phenomena. Likewise, we easily step into this kind situation or momentum after the honeymoon period we are in from a relationship, a marriage and a job, a position or an organization.

After you fully understand the reason why those questions are constructed, I believe you have fully geared up to conduct this Giant Story Questionnaire. So, go rock-n-roll.

5. What would you do if you have stood on the giant for too long?

Dealing with inertia – My intention of this question is to create awareness for our personal and work life. As we know inertia is a common phenomenon that happens to a person, to a couple, to a marriage or to an organization. In order for us to better understand "Inertia" we'd better look into the meaning of this word:

Inertia

noun.

1. the state of being inert; disinclination to move or act

2. (physics)

a. the tendency of a body to preserve its state of rest or

giant." (I followed with a question with this answer: What if another giant doesn't let you climb up?)

• "I would ask my giant to speak to another giant. When they become friends, I can ask my giant to ask another giant to lay his arm on my giant so that I can climb over." (This answer showed strong sense of social intelligence on the participant. I personally favored this answer very much since it was from my ten year old son, Jonah.)

• "I would ask my giant, do you know another giant over there and if he is dependable? If the answer is yes, I may climb over to another giant."

Just a few examples, I guess you may have already started to read the meanings between the lines. Questioning is a very intricate and interesting tool. It not only helps us understand others but also brings awareness toward our own sense of being. I believe learning is a two way process. Each time I conduct this questionnaire to others I have been learning non-stop.

4. If you see another giant walking by who is much taller than the one you are standing on, what would you do?

Drive – Ambitions and longing for growth. I originally asked this question to detect the participants' willingness to step out of their

comfort zone to grow and expand themselves. In fact, many of the responses were very intriguing. For example:

• "Wow, another big giant. He is taller and bigger than mine, but I am not sure I can trust him. I am just okay with my own giant."

• "I will jump off my giant and climb up on another

leadership characteristics mentioned in the above research done by Kouzes and Posner. Imagine if all the participants consciously choose to align their values with their feelings, thoughts, behaviors and actions. What kind of leaders would they be able to grow into and what kind of impact would they be able to bring to the world?

3. If the giant you are standing on fell, what would you do?

Change management – People tend to be heavily influenced by their surrounding environment. This question is to help people be aware of the initial reaction toward the precipitate event. Hopefully this question will help the participant come up with the constructive options that they are able to choose when others or the environment lets them down or disappoints them. Be cautious to ask this question since it may touch some sensitive part of the participant. I highly suggest to especially keep the responses of this question confidential.

Continues the table

Age	Sex	Profession	Question 2 answers
20+	Male	HR	Kind-hearted, generous and humorous
20+	Male	High Ed	Strong, supportive, larger than life, comforting
30+	Male	High Ed	Positive, empowering, authentic and caring
30+	Male	Pharmaceuticals	Confident, non-wavering, still, proud and able

Frankly speaking, when I first gathered all the answers for this question on the spreadsheet, I was so amazed at all the giants' personal characteristics. They are all so beautiful, uplifting and positive in a sense. I definitely had an Aha moment. I found myself somehow blinded toward all the beautiful people around me. If I didn't ask this giant question to them, I would never gain this awareness towards their beautiful inner values. It is crystal clear that almost every one of the participants possessed some of the

Continues the table

Age	Sex	Profession	Question 2 answers
40+	Female	Account Manager	Strength, protective, trustworthy
30+	Female	Teacher	Loving, stable, strong, honest, knowledgeable
40+	Female	Researcher	Brilliant, Learned and solid, grounded
30+	Male	Business Manager	Supportive, skillful, balanced and conscientious
20+	Male	Student	Dependable, trustworthy and strong
20+	Male	Student	Knowledgeable/wise, responsible, forgiving, loyal, kind
20+	Male	Student	Steady, trustworthy, companionship
30+	Male	High Ed	Experienced, caring, perspective
30+	Male	Coffee trader	Caring, driven, loyal
40+	Male	Professor	Gentle, generous and helpful
40+	Male	Professor	Powerful, tolerance, height, balanced

What personal characteristics does your giant have?

Age	Sex	Profession	Question 2 answers
20+	Female	Student	Kind, perseverant and faithful
20+	Female	Student	Leading, caring, wise and humorous
20+	Female	Student	Mature, strict/disciplined, caring
20+	Female	Student	Intimate, dependable
30+	Female	Interpreter	Courageous, dependable, caring
30+	Female	Lawyer	Dependable, secure, and logical
40+	Female	High Ed	Helpful, good follower; supportive
40+	Female	High Ed	Kind, gentle, compassionate, strong, humble
40+	Female	Lawyer	Reliable, high latitude, visionary
40+	Female	Student	Wise, kind, enlightened, seasoned and divine
50+	Female	Retired	Powerful, trustworthy, dependable/accountable
40+	Female	High Ed	Calm, humble, strong, selfless, thoughtful
50+	Female	R.N.	Gentle, patience, strength
40+	Female	Executive Asst.	Dependable, patient, insightful

encourage people to harness and expand their value repertoire.

James Kouzes and Barry Posner have spent more than twenty-five years on the study and research of leadership. In one of the studies they claim that over seventy-five thousand responses over many years yielded a list of characteristics that included, "honesty, forward-looking, inspiring, intelligent, fair-minded, broad-minded, supportive, straightforward, dependable, cooperative, determined, imaginative, ambitious, courageous, caring, mature, loyal, self-controlled, and independent." (as cited in Wildflower & Brennan, 2011, p.149). Interestingly enough, with the limited data I gathered from question two of the giant's questionnaire I found amazing correlations with Kouzes and Posner's findings. The following chart will present the small amount of responses to question two that I gathered for your reference.

people as you can. I guarantee you that their answers will evoke something in you. This question exercises people's envisioning muscle. Most importantly, it encourages people to exercise describing and communicating their vision to others.

2. What are the personal characteristics you see in this giant?

Detect and explore unconscious Values – This question is to reverse the perspective of the giant and zoom back into the giant's inner world. It helps us to see the participant's own inner values, which is projected onto the giant they stand on. These values are just the initial or raw values people have. The purpose of this question is to create awareness toward their own values and eventually

Luke Sun
2015

Chapter 7: The Giant's Story Questionnaire (Breakaway - Transformation) 59

Questionnaire with you. After that this questionnaire is going to be yours, it means you are free to conduct this questionnaire with anyone you want. I believe as long as you are willing to spend 5-10 minutes of your time to interactively conduct this questionnaire with the people around you, magic will happen each time you are doing this. If you are not sure, please make sure to give it a try.

1. Imagine you're standing on the shoulder of a giant, what do you see and how do feel?

Visioning exercise—It is a miracle question to engage people to zoom out of the limited self to see the world from above. It will give us a better perspective to see  the big picture while becoming aware of our smallness. I encourage you to conduct this question to as many

mentor or coach in their lives. I highly suggest reading his book since it is absolutely congruent with the concept of giants that I am trying to communicate in my book.

Evidently, giants are like leaders, mentors and coaches who lift up others. Giants are like positive forces (such as culture legacies, favorable environment) that push the world to move forward. Many times those giants appear in our lives as role models, mentors or coaches or a favorable environment or culture heritage in different periods of time. Eventually we are called to grow into giants who are able to offer our own shoulders for others to stand on. We are destined to grow up to lead in one way or another. Sooner or later we will be bound to lead the way for those who view us as giants.

In order to be well equipped with this practical concept and for us to effectively apply these concepts to our daily life, I would like to briefly review the Giant Story

Greco's storytelling class came along and here I am writing this book.

By now, you must have become acquainted with the metaphor of giants. It definitely is not a new concept. Confucius once said: "Out of three, there must be one who can teach me." In ancient China, a teacher is considered as a father, mentor or coach. Confucius perceptibly and discernibly portrayed that "giants" are often among us and with a humble heart we can definitely recognize and learn from those who are wiser around us. Just recently I was communicating the concept of the giant to one of my coworkers. After she heard my thoughts she immediately recommended to me to read Malcolm Gladwell's book Outlier (Gladwell, 2008). In his book he eloquently presented abundant examples of those who are successful, from Bill Gates, Steve Jobs to the Jewish community and his own success. These successes were largely because of their family background, culture legacies, great parenting,

Since then a few questions have been lingering in me, "How can I tap into people's unconscious heart in a short time? What do I need to improve to be more influential while coaching them? What are the gaps between me and those students?" These questions keep me hungry for knowledge and curious for solutions. I tried to recall the hundreds of students I talked to in the past. There is a definite common theme in most of them. They were all somehow disappointed with an authority figure and society (very much like myself when I was younger). I resonated with them. It was just a few months ago that I started asking a series of interactive questions in order to test or experiment ways of detecting another person's value. I believe what people value the most will determine their vision, goals and their behavioral pattern. At the beginning, I just want to know who is the giant that those students admire. But the more I asked the more surprised I was to find out people's responses. Then, Professor Janet

Chapter 7:

The Giant's Story Questionnaire
(Breakaway - Transformation)

Last year, I was introduced by one of my American friends, to some overseas Chinese students who were studying in Philadelphia. Through networking I gathered 6-8 students who were interested in the leadership group that I initiated (a combination of students from Drexel, Temple, University of Pennsylvanian and Saint Joseph University, one American born Chinese, one Caucasian and rest of them were Chinese overseas students). This group I led became side tracked because of the apparent irrelevant content and time constraint of all of us. I didn't take it as a failure, but a great learning opportunity.

Application Exercise

• When was last time you took a person or a job for granted? What made you appreciate the person or the job again?

• Can you name some of the inertias you walked into in your life? How did you became aware of the inertia situation and eventually overcome the inertia?

Luke Sun
2015

was he able to identify many giants in his life, but also he is a brilliant and humbled learner. By talking to him, it definitely humbled me. I suddenly started to identify the giants I have met in my life. A strong sense of gratitude overwhelmed me! Looking back, all those who played "giant" roles in my life have raised me up to be the person I am today. Reflecting deeper, I have to thank those who lifted up my giants' giants. Likewise, the country, the world and the universe that we are dwelling in are all the giants who made our life possible. There is a strong sense of connectedness and oneness felt by me for whoever is the ultimate Giant behind the whole universe. I must say that this song, You Raise me Up, explicitly speaks my mind "... You raise me up so I can stand on mountains / You raise me up to walk on stormy seas / I am strong when I am on its shoulders / You raise me up to more than I can be." (5-8).

the honor to work and learn from some of the best minds of the world. No matter how great a job is we are bound to cross over the honeymoon period. That is when the inertia hit me. After working at Wharton for about two years I started taking for granted all the great professors and the privilege of working at Wharton like everybody else. Until one day I asked one question to one of Wharton's faculty, James Thompson, "Knowing that you can see far away from the shoulder of the giant and you trusted the giant, what are the personal characteristics of the giant you are standing on?" He didn't answer my question directly, "Jack, I don't think this 'giant' is supposed to be singular. It should be plural, giants." At that moment I had an epiphany. I inhaled deeply and responded, "Aha, you are surely one of my giants!" We both laughed. I suddenly realized this prestigious professor didn't become eminent by just standing on one giant. He has been standing on many of the giants he met in his life. Not only

economic change. From the first person's experience I had witnessed how this change had violently tortured and distorted fragile human souls. Like many Chinese, I have been deeply concerned about my mother country's future, sustainability and her wellbeing.

I came back to American for the sake of my children's education because we couldn't afford to stay at the international school in Beijing after they terminated our scholarship for an uncontrollable reason. As the head of the family, I decided to move back to the States rather than moving to another city in China. I also realized that I needed to start a career for my family. With my unwavering passion for higher education I applied to all the colleges on the east coast. Fortunately, I accepted a position at the Wharton School as a program coordinator.

Not soon after I started working at Wharton I realized I am once again standing on the shoulder of another giant. It is a privilege to work at Wharton School since I have

Chapter 6:

Wharton Inertia
(Awareness - Society and beyond)

While reading a blog, I was struck to the core, when I came upon a quote that was rewritten by the NY Times: "China, please stop your flying pace, wait for your people, wait for your soul, wait for your morality, wait for your conscience! Don't let the train run out off track, don't let the bridges collapse, don't let the roads become traps, don't let houses become ruins. Walk slowly, allowing every life to have freedom and dignity. No one should be left behind by our era." (NY times, 2011). As much as I loved my students I found myself utterly powerless in front of the huge wave of moral degradation and drastic social and

Application Exercise

• When you find yourself helpless and despaired whom do you run to for help? How did you break away from your bottleneck situation?

• Imagine if you have a giant who will always be there for you and will never let you down, what would that experience look like for you?

for knowledge. I realized how limited and powerless I was in front of all the people that I cared and wanted to help. It was then that I learned to ask for emotional and spiritual support from our family and close friends in America, because I simply just couldn't do it alone.

During those three years of being so closely tied with hundreds of "Rich Second Generation" students, I had heard so many heart breaking stories, rarely joyful ones. I was compelled to lean on any stories that could give meaning to life for help, such as, many of the stories from the Bible. It was those stories that sustained my faith and hope in the midst of desperation and hopeless.

teaching plans three to four times each semester based on every two to three weeks of constant feedback from my students. Since I lived right on the college campus, each day after 8pm I was open to meet students in small groups, sometimes one on one for mentoring or coaching. Supported by my lovely wife, she knew how much those students needed our help. She decided after 8pm was the best time for me to meet my students since our kids went to bed at 8pm each day. I had tried whatever creativity I could use to motivate, to teach and coach/mentor them. There were times I had to counsel someone until after midnight. There were times I had to fly to another city immediately to do some crisis intervention. There were times I found myself utterly powerless in helping students who were battling depression and other psychological problems. There were times I had to surrender myself to a higher power for strength and wisdom.... All of those experiences had made me desperate for growth and hungry

"Rich Second Generation" since it had an obvious negative connotation attached. I found out that what they were going through somehow paralleled my college life. I was labeled as a "Farmer". Although "Rich Second Generation" and "Farmer" are at far opposite ends of the economic spectrum, interestingly these labels equally diminished who we really are. We are more than the labels the society and others had put upon us.

Through that untypical connection between my students and I, I found that there were no boundaries among us because deep down we all longed for respect for our soul and significance for our lives.

I was lucky to be supported by my department director who was in charge of me. She said to me, "I knew you have studied and worked abroad. I trust your ability. So, please do whatever you can do to make those students learn and grow." In the next three years, I composed my own teaching plans for each semester. I revised my

"What are you guys afraid of? Do I look like a tiger that will bite you when you ask questions? I do feel like I don't know you at all. I have to admit I won't be a good teacher if I don't even know my students. I am so glad we are finally talking now."

That was in 2009 when I almost gave up on teaching. After that class I scheduled with all of my students one on one free consultation or conversation to get to know all of them. It was like a whole new world was discovered. I had about 150 students for that semester. They were given the choice to come to talk to me in pairs or alone. Two worlds merged when we started conversations with each other. After many one on one and knee to knee conversations I learned that my students were a group of students who were born in the age of the economic bloom. Their parents were either rich businessman or government officials. "Rich Second Generation" was the name the society gave to them. However, almost all of them did not like the title,

tell me one reason I should be here teaching you guys."
I finally confessed to my students about my dilemma,
"When I gave out homework, none of you did your home
work. When I begged you to respond to my questions,
none of you participated. When I encouraged you to ask
questions during classes, none of you spoke up... I feel that
I am such a failure as a teacher who is unable to engage my
students." "No, teacher, we love your class. You are one of
the best teachers we had," one student raised his voice. "I
am not understanding it. If you like my class why don't I
see any engagement in the class, why don't you ever ask me
any questions or participate?" I uttered out with perplex.
There was 30 seconds of silence, but it felt like an hour.
"Teacher, we have many questions that we want to ask
you, I guess we are just too afraid to ask in public." One of
the students mumbled, "We are just not confident that we
are able to learn English well in two years since most of our
English language foundation was too poor to go abroad."

Chapter 5:

Crying for a Giant
(Need to grow- Experimenting)

It was a Friday morning, I rushed back to my 11am class.
I was 30 minutes late, as expected. Before I stepped into
the classroom I signaled to my wife (who was covering
the class for me) from the front door that I was back.
She announced to the whole class that their teacher was
back and said goodbye. After my substitute teacher, my
wife, left I saw the obvious confusion among my students.
Finally one of students asked, "Why are you in a suit,
teacher?" "I went to a job interview," I responded. "Do you
mean that you are not going to teach us anymore!?" There
was clearly a surprised look on that student's face. "Well,

• If this giant/organization lets you down or makes you disappointed how you would react and what would you do?

Luke sun
2015

leaves UPS, the HR department would close the file and the employee cannot be hired again. But, he said UPS would keep my status open, which means if I want to come back to UPS in the future, the door will stay open for me. I have to admit even after I left UPS for years I still feel part of me still bleeding "Brown". The legendary company taught me a legendary lesson about leadership. There are not enough words to express my gratitude toward those great leaders that I knew at UPS.

Application Exercise

• If you view the organization you are working at as a giant, what kind of personal characteristics does this giant have?

this meeting?" I hesitated. "Come on, Jack, we know how capable you are, and we absolutely trust you!" He looked at me with a firm and uplifting trust I couldn't resist. Up until today I don't even know how and where all the courage came from for the meeting. I not only led the meeting with great professionalism but also helped our team secure my first million dollar contract at UPS.

Due to my children's education, my wife and I decided to move back to China so that they are able to learn Chinese. I said goodbye to UPS toward the end of 2008. I still couldn't forget the day I said goodbye to my coworkers from the business development department. My director presented a plaque to honor me for serving at UPS for one and a half years. She asked me to give an account for where I was heading with my life journey for everyone to hear in my department. It was very ceremonial that I felt it put a wonderful end toward my journey at UPS. After that, our division director told me that in general after an employee

brand with the utmost professionalism. It was because I believed that I was the UPS brand my director told me that I was. UPS was one of the best organizational cultures I have ever experienced. The director who told me those words was just one of the many great leaders I had met at UPS. I remember another manager who was always encouraging and trusting. Following a lead, I had secured a meeting with one of the biggest electric equipment supplier companies on the east coast. In that meeting, I arranged two of our UPS logistics experts and one of my managers to meet the company's CEO, CFO, lawyer and Chief Logistic Officer. Oddly enough, five minutes before the meeting, my manager said to me, "Jack, it was because of your great effort that we are able to rebuild our business relationship with this customer. I understand that you know them much better than we do. Would you mind taking the lead in the meeting?" I looked at him and responded with doubts, "You believe that I am able to nail

jobs I had ever had so far. Not to mention my language barrier on top of learning all the technology and corporate knowledge and culture, to sell UPS to all kinds of demanding customers was truly a challenge even for a native speaker. It was these experiences, however, that pushed boundaries and propelled me to grow drastically during the time I served at UPS. UPS would be considered a giant in every aspect of its business: Leadership, culture, industry, legend, globalization, technology, and so on. I was like a sponge soaked in all the knowledge from the first day I started working at UPS. After a few weeks of training, one of my directors told me, "Jack, you are going to be the brand of UPS, and you are going to represent UPS in front of our customers. So, go rock!" Those words were so empowering that I felt like a superman regardless of all the adversity I had faced. Looking back, when facing all the cold calls, unhappy, and sometimes angry and mean customers, I was able to present the best "UPS"

Chapter 4:

UPS Story
(Legendary Leadership-Organization)

It's not hard to draw the conclusion that my mother, my German mentor, my wife and my principal friend are all the giants I have met in my life who have more or less influenced and shaped my value and character, as well as my leadership capacity. It is these episodes of my life stories that I chose to remember. They have formed the person that I am today. I now clearly see these collective qualities that I have learned from them: forgiveness, honesty and truthfulness, trusting, and empowering.

When I was hired by UPS as a sales representative back in 2006, it may have been one of most challenging

Application Exercise

• Identify the giants you have met through your life. Write their names and the personal characteristics that made them giants.

• Write down the moments of how those giants have impacted you on a personal level.

will feel so relieved to have me in America now that I am able to survive with my language skills. For that I am very thankful for all of your hard work." It was so sweet to hear those words. All those times I had doubts about staying there without being paid. It seemed so pointless and insignificant at that moment.

The principal was the one who talked to me about standing on the shoulder of the giant. Through the experience she and I have become friends who forget our age differences. Looking back she is surely one of the giants I met in my life who has lifted me up with her teaching ethic. It was after that experience that I started to realize that there are all kinds of leaders. It is up to us to decide what kind of leader we choose to be. Like my principal friend, who chose to be a leader who empowers others even when the cost sacrifices her own interests.

kindest lady I have ever known. She said to us, "I am no businessman, and I know I can't make the CEO change his mind about closing the center, but I am an educator whose intention is only to disseminate knowledge to my students and help them to excel. I believe that you all believe what I believe. As long as there is one student here I will be here to help this student until the last minute."

It was maybe one of the noblest times in my life. Without being paid for two months, I and the principal, and a few other employees stayed at the training center literally until the last minute. It was on that day that one of the many students who had truly benefited from our training center said to us with tears in her eyes, "I am 55 years old and never thought I would be able to learn English. It was your training center that gave me hope and helped me gain my confidence to learn English. I have learned how to carry on daily conversations after only five months of learning here. I know my daughter

Chapter 3:

"On the Shoulder of a Giant"
(Leadership drama - Group)

I got my first official job in an English training center soon after I moved to Beijing. I worked at that training center for about seven months until it was closed. Two months before the training center was going to be forced to be shut down, the CEO had stopped paying his employees. As a result, most of the employees left for new opportunities. Some employees took away some of the company's expensive property and justified their actions because they were not being paid. There were about three or four employees left working at the center. One of them was the principal of the training center. She was the

logic is this?!" My wife, as my life partner, basically pushed me from the side with those words and encouraged me to step up not only as a man but also as a leader for our family. It was that trust that propelled me to be a better man and leader everyday.

Application Exercise

• Was there any moments in your life that you found yourself so empowered by someone? If so, write down your experience and share your experience with someone who you care about.

• Please watch the Youtube video on leadership, "First Follower Leadership Lessons from Dancing Guy" and reflect on what makes a leader a leader: https://www.youtube.com/watch?v=1euTzm4-pMU

she was born as a crying baby, when she was six or seven years old with happy smiles on her face, here and now a gorgeous and lovely angel standing right in front of me, and finally as an old lady with grey hair but still having the glowing loving look in her eyes. That moment I felt a sense of serenity and peace I never had before. "This is it. She is the one." Fourteen years later we are still writing this beautiful and romantic story together. I can't thank God enough because He sent to me such a wonderful woman. One sentence that my wife said to me in our early marriage continues to ring a bell in my mind: "Jack, you are the head of our family. I will follow you, and so will our children in the future. I just want to let you know that your decisions will be our decisions. Even if I don't always agree with the decision, as long as you insist, we will still follow you." I remember I said to myself when I first heard her telling me this, "What, even if you do not agree with the decision, you and our kids are still going to follow me? What kind of

many times what a great leader you were when you were our monitor in college because you truly cared about us and were always positive and empowering." It was truly surprising to me how my classmates would picture me like that, especially if she knew what I was really like the first two years of college. Anyway, it did make me realize how my mentor's words had drastically altered my life story and even affected the stories of those people around me. This may have been my first glance at what my giant (my mentor) had lifted me up to be.

Romance was never a word I fully understood until I met my now wife at the end of my senior year. I didn't fall in love at first sight. However, I did at second sight. The memory is still as fresh as an HDTV. She was waiting for me with her Dad to play basketball at the school's courtyard. That 's the second time I saw her. I am not sure what the difference was from the day before when I saw her in one of the classrooms on campus. But the second I saw her, her whole life flashed through my mind: when

Chapter 2:

Spark of Romance
(Unexpected Push from the Side - Pair)

I had been mentored by my German friend for about one and half years. As a result, even the darkest clouds wouldn't stop me from seeing the blue sky. Thanks to my great mentor, I not only came out of my depression, but I also matured emotionally and spiritually as a whole person. I guess I had successfully made it into the self- authoring mind (Kegan & Lahey, 2009) with his help. To be honest, I was utterly happy with being just who I was regardless of the fact that I was still a lonely and poor countryside boy in my senior year. Another outcome was told by one of my classmates years later which I was totally not aware of at that time. She said to me, : "I have told my students so

This is my heavenly treasure

Application exercise

• Was there any moment of your life you found hopeless? Who or what guided you through those difficult times?

• What are the heavenly treasures you hold dear to your heart?

foreigners I had known before. As an atheist at the time I started to be suspicious that the "human does have a soul." This German Bahai's story obviously fascinated me with the wildest imagination about the mystery of the universe. Little did I know he was going to be one of the greatest mentors of my life. One day, after our language practice session, we sat in the dawn of the early fall on campus. I gathered all of my courage and told him all of my struggles toward life, even my darkest secrets. After I confessed my dirty laundry, I was so afraid he was not going to want to be my friend anymore. What he said next to me became a story line I told again and again to my friends in my life. **"Jack, honesty, integrity, kindness and truthfulness are things nobody can rob away from you unless you give them up on your own. They are heavenly treasures that will stay with you forever if you choose to."**

studying. Darkness was the only thing I remembered for the first two years of college. No matter how I struggled to hold myself together, I saw no hope, no meaning. I was hopelessly and desperately lost. Every day I tried to put a rigid smile on my face and pretended that I was okay. However, the turmoil of my heart was so heavy that I didn't even know who I was anymore. It literally felt like I was stuck in the darkest tunnel and yelling for help but no one was there to listen.

In my sophomore year, as an English language student, my study load became heavier than ever. It temporarily kept me away from my restless inner world. It was like a ticking bomb that could blow up at any time. By a very random chance, I got to know a German friend who was studying Chinese at the college next door. In the beginning it was like give and take: he helped me practice English, and I helped him practice Chinese. Gradually, I came to realize this German didn't fit into any of the categories of

standing right in front of them. "That's farmer mentality," "You are so farmer," etc.... They may have been half joking, but all farmers' offspring knew that we were definitely being discriminated against because of our cultural heritage and economic background. I even knew a few students from the countryside who pretended that they were from the city to fit into the popular crowds. Gradually, I started struggling with my identity. Some of the classmates told me I was too honest and smiled too much. They told me if I stayed cool I would draw attention from girls. I found myself lost in transition. For a while I tried to be cool, manipulated with lies, and not to reveal the soft side of me. I was depressed and engaged in destructive behaviors like alcohol, porn and suicidal thoughts. During the time, I tried to pursue a few girls around me. They all ended up as the stereotypical story: a hopeless poor farmer boy wants to date a rich city girl, and it just simply didn't work. In the midst of the chaos of my inner world, I couldn't focus on

take more than we give because that way we can live a life with peace. That day I held my mother's words dear to my heart. That was one of best lessons on "Give and Take" (Grant, 2013) my Mother has given to me.

After seeing my parents pulling their teeth and borrowing just enough for me to go to school, I secretively made a promise to myself that I would study hard and get a good job to repay the debts my parents owned. With that intent in mind, I stepped into the mysterious world of college.

It didn't take too long for me to figure out that college wasn't a romantic comedy like I had imagined. First, I realized that as a countryside boy trying to fit in was extremely difficult, not to mention the drastically different lifestyle of the city people who lived around me. The hardest part was knowing how a farmer was viewed in their eyes. Very often I would hear city students make fun of country people even though they knew I was a country boy

borrow money from him. The uncle we visited was one of my Mother's younger brothers. He could help me pay for the whole school if he wanted. I don't remember what my mother said to my uncle, I just remember what my uncle said, "Here is 200 Yuan ($25). You can take it and you do not need to return it to us." My Mom took the money with an uneasiness that I had never seen before. She rigidly pulled herself together and signaled me to leave. The second she walked out of my uncle's door I saw her eyes well up with tears that slid down one side of her face. I couldn't imagine how hard it was for her since my parents had helped my uncle so much when they were poor back in the old days. As the oldest of her eight siblings, my Mother was forced to drop out of school when she was in 5th grade so she could take care of her younger siblings. She said to me later that day that she wasn't a good older sister because she hadn't taught her brother good values when he was younger. She also told me that we should never

colleges, for most high school students of the late 90's going to college was like winning the lottery. I somehow scrambled through the gate of college with "pure luck" according to the exact words of one of my teachers. I remembered clearly the day I was accepted into college. The whole village seemed to treat me like a celebrity. Before that happened, when I walked on the street, no one would even notice me. Suddenly, they looked at me with admiration and jealousy. It was like I had overcome the whole world.

Walking on the clouds didn't last for too long as the time to register for college drew closer. At the time, most farmers did not have much savings, had no credentials and didn't even have the knowledge to borrow money from the bank. The only way for my parents to get enough money to pay for my tuition was to borrow from relatives and friends. I will never forget the day my Mom walked out of one of my uncle's house when she took me with her to

Chapter 1:

The Darkest Moment of My Life
(Light in the Tunnel – Individual)

Those born in the 80's in mainland China were considered a very fortunate generation. It was the time China recovered from World War II, the civil war, and the '60's famine. The message I received from the mainstream media was that China (the Sleeping Giant) is waking up from its sleep. Like the rest of the general public, I somehow had the feeling that we would be taken care of by this "awakening giant." At least, the news and TV told us so. However, when it came to my personal story it spoke otherwise.

Because of a 3-7% acceptance rate to the state-owned

is beyond what the single word "dream" can depict. If I summarize in a sentence my life up until today, I would call it a capricious leadership journey.

Back in 2003, before I headed to the United States to live and study, one of my coworkers said one sentence to me, "Jack, you are going to stand on the shoulder of a giant and make sure to take full advantage of this rare opportunity." I couldn't fully comprehend the sentence at the time, but it has stuck with me for the rest of my life.

Background

I grew up in a rural area in the middle of China in the place where the Terra Cotta Army gained its worldwide fame. I was one of the many less privileged farmer's offspring who endured an impoverished and harsh life. As I close my eyes even now, I can still feel the icy cold stone bench I sat on in the freezing cold classroom with broken windows covered by paper-thin plastic. That was one of the permanent memories I received from the first winter in my hometown primary school. Fate has its own way: from that meager beginning through many serendipitous life events, I am now living in America and working at one of the most prestigious business schools in the world, The Wharton School. It sounds like I have pretty much achieved the American dream. I admit that my life story

5. What would happen if you were standing on the giant for too long?

Now that you have completed the Giant Story Questionnaire, I am pleased to welcome you to my personal journey discovering my own giant. I sincerely hope this journey will not only create some awareness, evoke some thoughts, but also inspire you to be ready to take some impactful actions.

Luke San
2015

3. If the giant you are standing on fell, what would you do? What if he may not be able to stand back up again? Write down what you would do after the giant fell.

4. If you see another giant walking by who is much taller than the one you are standing on, what would you do?

1. Imagine you are standing on the shoulder of a giant. What would you see and how would you feel? Use your imagination. This giant can be as tall as you can imagine. Now write down or draw your answers. Whichever makes the most sense to you.

2. Since you are standing on the shoulder of this giant, it means there must be some level of trust between you and the giant. So, list 3-5 adjectives to describe what the giant's personal characteristics are?

The Giant Story Questionnaire

Through the use of the Giant Story Questionnaire, I am inviting every reader to be an active participant in this ongoing research. Through this hands on experience, I hope you will sense the magic that this line of questioning will bring to you and your world. Meanwhile, I have intertwined the Giant Story Questionnaire with some extra exercises throughout the book to further your practice in a more productive and profound manner, if you wish.

Before you continue reading this book, first finish the following questionnaire as this will help you to better engage with the rest of the book.

are measured qualitatively – measured by our heart.

Giant

"Giant is the English word (coined 1297) commonly used for the monsters of human appearance but prodigious size and strength common in the mythology and legends of many different cultures. The word Giant was derived from one of the most famed examples: the gigantes of Greek. (Wikipedia)"

There are also the giant of Goliath from the Old Testament, and the giant from the fairy tales of "Jack the Giant Killer." While the list is going on and on. In this book, I have used the word "giant" to embody the intangible soft values. So, the "giants" mentioned in the book are very different from the "giants" to which Wikipedia refers. Isaac Newton once said, "If I have been able to see further, it was only because I stood on the shoulders of giants." This book invites all readers to embrace Newton's definition and join in the pursuit of their own unique giants.

Webster)

The value or values I mention in this book, however, are different from the economic value we refer to in general. On a Chinese online Wikipedia website, value created by an organization or enterprise was divided into two kinds: Soft Value and Hard Value. Hard values are tangible values, such as: commodities, equipment, factory building and other visible, touchable and concrete things. Soft Values are registered trademarks, technology, corporate image, customer satisfaction and other invisible and intangible things (Baike).

The Soft Values mentioned above are under the lens of the business world. The values mentioned in this book are more about soft values on a personal level. in other words, soft values are a kind of intangible quality one possesses. Einstein once said, "Not everything that counts can be counted, and not everything that can be counted counts." What Einstein was trying to say is that the Soft Values cannot be measured quantitatively. Instead, these values

Introduction

In this book, combined with the frontiers of knowledge from psychology, social psychology, organizational dynamics and leadership research, I weaved a series of my own life stories with an interactive survey into the narrative. It will compel and inspire the readers to seek the revelation of their own unique inner giants. Before reading this book, a clear definition of the following words will definitely help all readers better understand the intention of the author.

Value

According to Merriam-Webster online dictionary, value is: the amount of money that something is worth: the price or cost of something; something that can be bought for a low or fair price; usefulness or importance (Merriam-

that stories have their positive sides too. Doctor Janet Greco's beautiful vision on stories gradually awakened my innate self-authoring heart. I must say that her vision has evidently altered the trajectory of my life story. As a result, I collected some of my valuable life moments into a story form. I hope these stories will evoke your desire to seek your own stories that are unique and instrumental to your life too.

chart. Each one of us has a story, likewise, each group and each organization has its own story. No matter what genre, fiction or non-fiction, it matters only as long as they are aesthetically used as an instrumental tool. These stories will positively impact individuals, groups, organizations, and our society and beyond as a whole. Following Janet Greco's vision I have organized my book in a similar fashion.

Before I took Professor Janet Greco's "Stories in Organizations" class, I was a total anti-story kind of guy. Because subconsciously related stories with rumors and gossip or something untruthful, I grew to dislike the word "story". Sometimes, when I saw my wife weeping while reading a novel or watching a movie, I couldn't help saying, "I really don't understand why you are wasting your precious time and emotions to cry about something that is not even real." Right after Doctor Janet Greco presented her beautiful vision in front of me, I started to realize

program. Her teaching has inspired me to collect all of my giants' stories. I would not have written "Seeking the Giant" if it weren't for her beautiful vision from her very first class. I sketched and revised her vision into the following drawing to share with you:

Professor Greco's vision about stories is systemic, instrumental and beautifully incorporated in this diamond

my book. Thanks also to Joern Geisselmann, Wei Fang, Margaret Baumann, Edmond Baumann, Tony Heath, Karen Hopkins, Rebekah Zanders, Steve Zanders, James Thomason, Todd Henshaw, Sue Wharton, Paul Wharton, Deb Giffen, Sumathi Pearl, Adam Grant, Anne Corcoran, John Percival, Mike Useem, and all those who have left a mark on my heart.

I also must admit I am greatly indebted to the free Library of Philadelphia for the audio book application called Overdrive. Because of the free library and the modern technology I am able to access countless great audio books on leadership.

Finally, I have to give my exceptional thanks to my professor from the University of Pennsylvania, Janet Greco, whose class from the Organizational Dynamic Masters Program: DYNM 673: Stories in Organizations: Tools for Executive Development is by far one of the most practical and beneficial classes I have taken in the

Acknowledgement

Special thanks to my beloved wife who has been there for me from the very beginning and supported me in the writing of this book. Her emotional support and editorial help has made everything possible. Thanks to my sons Jonah and Luke Sun who have sacrificed many of their play times with me.

Thanks to my parents who have raised me up to be the man that I am. Likewise, I am very thankful for my parents-in-law George and Carol Benner and their ongoing love and support for our family.

Great thanks to all those great teachers, mentors, and coaches who have influenced me in my life journey. And also, I wish to give my sincere gratitude toward Steve Shepard's encouraging words and great advice on

deeply indebted to your generosity and help. Because of you, I am able to stand tall, and I feel propelled to help and impact others.

I also want to dedicate this book to my sons, Jonah and Luke Sun. I hope this book will help you know me in a different light. May it be a good guidance for your life's journey.

Preface

This book is written first and foremost for my former students whom I taught at Renmin University of China from 2009 to 2012. I am here to thank all of you for being my students and for all the trust you have given to me. Without your trust and friendships, I wouldn't be inspired and empowered to write this book. I hope this book will connect you to your inner giant.

This book is also written for those who want to grow and break through their life norms and excel to lead. Last, but not least, this book is also written to encourage mentors, coaches and leaders to continue to impact our society sustainably and positively.

With insurmountable gratitude to all of my great mentors and coaches I have met through my life, I feel

It is through his own humble life story that Jack Sun teaches us about relationships with our own "giants." A compelling and interactive tale of one human journey to find meaning and significance, and a prescription for thinking about our lives, relationships and our impact on others. Giants grow future giants!

—**Todd Henshaw**: Director of Executive Leadership Programs, The Wharton School; President, Leader Development Associates Prior to Wharton, he was also the Director of Leadership Programs at the United States Military Academy at West Point

Endorsement

"There are few books that offer nuanced personal history as serious opportunities for others to learn actively. Jack Sun's valor and candor work their magic through narrative and metaphor, thereby generalizing his story for each of us to use as we wish. He is a giant whose shoulders can serve well, privately and professionally, for growth and increased happiness."

—**Janet Greco, Ph.D.** Faculty, Organizational Dynamics Masters Program, Wharton Executive Education, University of Pennsylvania

Contents

Chapter 5: Crying for a Giant
(Need to grow - Experimenting)　　42

Chapter 6: Wharton Inertia
(Awareness - Society and beyond)　　49

Chapter 7: A Giant's Story
(Breakaway-Transformation)　　54

Chapter 8: The Virtuous Circle of Life
(Application-Aesthetic and instrumental)　　72

Appendix　　83

Bibliography　　86

Contents

Endorsement	01
Preface	03
Acknowledgement	05
Introduction	10
The Giant Story Questionnaire	13
Background	17

Chapter 1: The Darkest Moment of My Life
       (Light in the Tunnel - Individual)   19

Chapter 2: Spark of Romance
       (Unexpected Push from the Side - Pair)   28

Chapter 3: "On the Shoulder of a Giant"
       (Leadership drama - Group)   32

Chapter 4: UPS Story
       (Legendary Leadership-Organization)   36

图书在版编目 (CIP) 数据

追寻巨人 ／（美）予森（Sun, J.）著译 . —北京：中央编译出版社，2016.1
（2016.2 重印）

ISBN 978-7-5117-2820-3

I.①追… Ⅱ.①予… Ⅲ.①成功心理－通俗读物 Ⅳ.① B848.4-49

中国版本图书馆 CIP 数据核字 (2015) 第 259825 号

追寻巨人

出 版 人：刘明清
出版统筹：董　巍
责任编辑：邓永标
责任印制：尹　珺
出版发行：中央编译出版社
地　　址：北京西城区车公庄大街乙 5 号鸿儒大厦 B 座 (100044)
电　　话：(010) 52612345（总编室）　　(010) 52612371（编辑室）
　　　　　(010) 52612316（发行部）　　(010) 52612317（网络销售）
　　　　　(010) 52612346（馆配部）　　(010) 66509618（读者服务部）
传　　真：(010) 66515838
经　　销：全国新华书店
印　　刷：北京时捷印刷有限公司
开　　本：787 毫米 ×1092 毫米　1/32
字　　数：120 千字
印　　张：5.75
版　　次：2016 年 2 月第 1 版第 2 次印刷
定　　价：28.00 元

网　　址：www.cctphome.com　　邮　　箱：cctp@cctphome.com
新浪微博：@ 中央编译出版社　　　　微　　信：中央编译出版社 (ID：cctphome)
淘宝店铺：中央编译出版社直销店 (http://shop108367160.taobao.com) (010)52612349

本社常年法律顾问：北京嘉润律师事务所律师　李敬伟　问小牛
凡有印装质量问题，本社负责调换，电话：010-55626985

Seeking the Giant
寻巨人

[美] Jack Sun (孙大) 著

中央编译出版社
Central Compilation & Translation Press